Field Hydrogeology

The Geological Field Guide Series

Field Hydrogeology, Fifth edition Rick Brassington

The Field Description of Metamorphic Rocks, Second edition Dougal Jerram and Mark Caddick

Drawing Geological Structures Jörn H. Kruhl

Basic Geological Mapping, Fifth edition Richard J. Lisle, Peter Brabham, and John W. Barnes

The Field Description of Igneous Rocks, Second edition Dougal Jerram and Nick Petford

Sedimentary Rocks in the Field, Fourth edition Maurice E. Tucker

Field Geophysics, Fourth edition John Milsom and Asger Eriksen

Field Hydrogeology

FIFTH EDITION

Rick Brassington
Consultant Hydrogeologist, Warrington, UK

This edition first published 2024

Edition History
McGraw-Hill. (1e, 2001); CRC Press. (2e, 2011); John Wiley & Sons Ltd (3e, 2007 and 4e, 2017)

Registered Office(s)
John Wiley & Sons, Inc., 111 River Street, Hoboken, NJ 07030, USA
John Wiley & Sons Ltd, The Atrium, Southern Gate, Chichester, West Sussex, PO19 8SQ, UK

For details of our global editorial offices, customer services, and more information about Wiley products visit us at www.wiley.com.

Wiley also publishes its books in a variety of electronic formats and by print-on-demand. Some content that appears in standard print versions of this book may not be available in other formats.

A catalogue record for this book is available from the Library of Congress

Paperback ISBN: 9781394180622; ePub ISBN: 9781394180646; ePDF ISBN: 9781394180639

Cover Image: © James Osmond/Getty Images
Cover design by Wiley

Set in 9.5/12.5pt TimesLTStd by Integra Software Services Pvt. Ltd, Pondicherry, India

CONTENTS

PREFACE TO THE FIFTH EDITION

I had resolved not to revise Field Hydrogeology after the third edition was published and then again after the fourth one was published and now I have done it again, and this time it was entirely my fault. I realised that there were a number of inadvertent errors in the fourth edition despite the great efforts that the Wiley system puts in to make sure that the book is correct and also my efforts. The reality is that even after months between writing something and reading it again it is still possible to read what you thought you had written rather than what is there.

I also wanted to put some additional things into the book which are small additional sections of text but which I think are important to include in a field manual. These included an example of a recent investigation of a flooded cellar and the difficulties caused by iron and manganese in clogging boreholes. There were a number of small changes but the biggest new addition covers flow through the vadose zone. This came about when I failed to explain this to intelligent people who had no knowledge of geology who decided that they would rather accept a simple picture of rapid recharge flow with the unrealistically high value for the hydraulic conductivity in the vadose zone rather than the complex system that nature has provided for us. You will have to read it to see what I mean; it is now at the end of Chapter 3.

The underlying theme of the book is the same in an attempt to provide a hands-on guide to field measurements in hydrogeology. It is primarily aimed at undergraduate students of geology, environmental studies, and civil engineering, as well as postgraduate students and people setting out in their careers as hydrogeologists and anyone who is interested in groundwater. I continue to believe in the fundamental importance of understanding the geology of an area in investigating its hydrogeology, and the importance of field measurements in developing this understanding. I hope that you will always make these the foundation of your work as they are crucial in gaining a reliable understanding of groundwater systems.

Rick Brassington
2023

PREFACE TO THE FOURTH EDITION

I was asked to revise my book by Rachael Ballard, my then Wiley editor, both to update it and to produce material for an e-book edition; this I hope I have done. As always, the book is intended as a hands-on guide to field methods in hydrogeology, explaining what techniques are needed and how to carry them out. The book's layout mirrors the logical sequence for the development of a conceptual model to understand the hydrogeology of an area, with the associated field studies needed to validate that understanding. Processing and interpreting the information collected during a data-gathering exercise is necessary to develop your conceptual understanding of a particular site and, consequently, these tasks have also been included. This fourth edition updates the information on the most recent field methods and instruments, although my practical approach includes suggestions on improvising measurements when specialist equipment is not available or cannot be used. I have included an additional case history, bringing the total to five, again using my projects as inspiration. They illustrate how field measurements are interpreted and the interrelationships between different aspects of groundwater systems, as well as how to take measurements in difficult situations.

Although the book is primarily aimed at graduate and undergraduate earth science students, earlier editions have also proved useful to many others. Consequently, I have rearranged the contents of Chapter 10, moving some to Chapter 1, and have included how to carry out a borehole prognosis, despite it being a desk-top study rather than fieldwork. There is a greater emphasis on abstraction systems, including those from shallow aquifers; I have included geothermal systems, which are growing in popularity; have rewritten the section on the impact of large excavations on groundwater systems; and included a short section on soakaway systems. However, the underlying theme of the book continues to be the fundamental importance of the geology of an area in trying to understand its hydrogeology, and the significance of field observations in developing this understanding. Both are essential for reliable hydrogeological interpretation, so make sure that they are always a central part of your work.

Rick Brassington
2016

ACKNOWLEDGEMENTS

A large number of people have provided helpful comments and guidance during the planning and preparation of this revision and I regret that it is not possible to name them all. My thanks are due to Paul Thomson on helping me get my thoughts together on flow through the vadose zone. In Wiley I would like to thank Sarah Higginbotham Senior Commissioning Editor who got the ball rolling Judy Howarth Senior Managing Editor who managed the project, Dr Frank Weinreich who sorted out many possible cover designs, and Durgadevi Shanmugasundaram for her help and patience in putting together my draft. I am also indebted to several of my clients for permission to use data from projects I have completed for them, including those that asked for the site details to remain anonymous.

Grateful acknowledgement is made for the following illustrative material used in this book as follows:

Figures 1.3, 4.14, and 8.1, and Table 6.5: CIWEM; Figures 2.2, 4.17, 4.18, 4.21, and 4.27, and Tables 2.1 (part), 6.5 (part), and 10.1: Environment Agency © Environment Agency; Figure 3.1a: Impact Test Equipment Ltd; Figures 3.4 and 7.8, and Table 3.1 (part): US Geological Survey; Table 3.2: Dept. of Economic and Social Affairs, *Groundwater in the Western Hemisphere*, United Nations (1976); Figure 3.6: redrawn from the *Groundwater Manual*, US Department of the Interior (1995) and used with permission of the Bureau of Reclamation; Figure 3.13a: Beckman Coulter Ltd; Table 4.1: Cambertronics Ltd; Figures 4.4 and 7.2: In-Situ Europe Ltd; Figure 4.15: redrawn from Hubbert (1940); Figures 4.18 and 7.5: British Geological Survey © NERC All rights reserved; Figures 6.16 and 10.1, and Table 3.1 (part): Geological Society of London; Figure 6.14: US Army Corps of Engineers; Figure 7.4: Solinst Canada Ltd; Figure 7.6: Isotech Laboratories; Figure 7.7: Department of Geology and Geophysics, University of Utah; Figure 9.4: European Geophysical Services Ltd; Figure 10.4: Springer-Verlag Berlin Heidelberg, Copyright © 2015; Figure 10.7: redrawn from Brassington (2014) by permission of Extractive Industry Geology Conferences Ltd; Box Figures 1.1 and 1.2 and the data for Case History 1: Shepherd Gilmour Environment Ltd; Box Figure 2.1, the data for Case History 2, and the example used in Section 10.5.1: Newcastle-under-Lyme Borough Council.

1
INTRODUCTION

Groundwater provides an important source of drinking water over much of the world. It also has the fundamental importance of maintaining river and stream flows during periods without rain and also supporting wetland sites. Groundwater is under threat worldwide from over abstraction and by contamination from a wide range of human activities. In many countries, activities that may impact on groundwater are regulated by government organisations, which frequently require hydrogeological investigations to assess the risks posed by new developments.

Pumping from new wells may reduce the quantities that can be pumped from others nearby, cause local spring flows to dwindle, or dry up wetlands. The hydrogeologist will be expected to make predictions on such effects and can only do so if he or she has a proper understanding of the local groundwater system based on adequate field observations. It is equally important to evaluate groundwater quality to ensure that it is suitable for drinking and for other uses. Groundwater commonly provides the flow path that allows pollutants to be leached from industrially contaminated sites, landfills, septic tanks, chemical storage areas and many more. Hydrogeological studies are needed to define groundwater systems in order to prevent such contamination or manage its clean-up. This book is concerned with the field techniques used by hydrogeologists to evaluate groundwater systems for any or all of these purposes and with the primary or initial interpretation of the data collected in the field.

1.1 Groundwater Systems

Groundwater is an integral part of the hydrological cycle, a complex system that circulates water over the whole planet; this is illustrated in Figure 1.1. The hydrological cycle starts as energy from the sun evaporates water from the oceans to form large cloud masses that are moved by the global wind system and, when conditions are right, precipitate as rain, snow, or hail. Some of it falls onto land and collects to form streams and rivers, which eventually flow back into the sea, from where the process starts all over again. Not all rainfall contributes to surface water flow, as some is returned to the atmosphere by evaporation from lakes and rivers, from soil moisture, and as transpiration from plants. Water that percolates through the soil to reach the water table becomes groundwater. In thick aquifers, groundwater at depth is below the depth of freshwater circulation and is saline, often with a higher electrical conductivity than seawater. The same is true for groundwater down dip from the outcrop of an aquifer. Groundwater flows through the rocks to discharge into either streams or rivers. In coastal areas, groundwater discharges into the sea, and the aquifer contains seawater at depth. The volume of water percolating into the aquifers defines the groundwater resources that both support natural systems and are available for long-term water supply development. In most groundwater studies it is necessary to consider the other components of the cycle as well as the groundwater itself in order to understand the groundwater system. Consequently, hydrogeological investigations usually include a range of field measurements to assess these parameters.

Groundwater flow through saturated rock is driven by a hydraulic gradient, which, in unconfined aquifers, is the water table. Rocks that both contain groundwater and allow water to flow through

Field Hydrogeology, Fifth Edition. Rick Brassington.

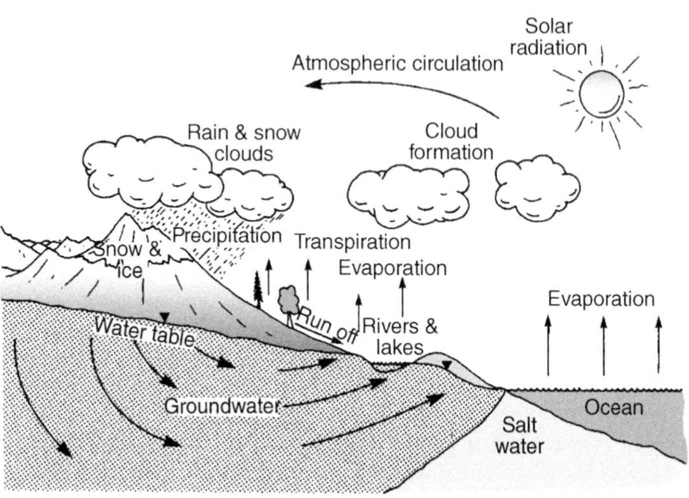

Figure 1.1 *The hydrological cycle.*

them in significant quantities are termed *aquifers*. Flow rates that are considered as significant will vary from place to place and also depend on how much water is needed. Water supplies to individual houses require small groundwater abstractions compared to wells supplying a whole town. In pollution studies, even small groundwater flow rates may transport considerable amounts of contaminant over long periods of time. A critical part of the definition is that the rock allows a flow of water, rather than simply containing groundwater. Some rocks such as clays have a relatively high water content, although water is unable to flow through them easily. Other rocks may not be saturated but still have the property to permit water to flow, and therefore should be regarded in the same way as an aquifer, a clear example being the part of an aquifer formation that lies above the water table.

Unless groundwater is removed by pumping from wells, it will flow through an aquifer towards natural discharge points. These comprise springs, seepages into streams and rivers, and discharges directly into the sea. The property of an aquifer that allows fluids to flow through it is termed *permeability*, and this is controlled largely by geological factors. Properties of the fluid are also important, and water permeability is more correctly called *hydraulic conductivity*. Hydrogeologists often think of hydraulic conductivity on a field scale in terms of an aquifer's *transmissivity*, which is the hydraulic conductivity multiplied by the effective saturated thickness of the aquifer, often taken being the depth of local wells.

In both sedimentary rocks and unconsolidated sediments, groundwater is contained in and moves through the pore spaces between individual grains. Fracture systems in solid rocks significantly increase the hydraulic conductivity of the rock mass. Indeed, in crystalline aquifers of all types, most groundwater flow takes place through fractures, and very little, if any, moves through the body of the rock itself. Some geological materials do not transmit groundwater at significant rates, while others only permit small quantities to flow through them. Such materials are termed *aquicludes* and *aquitards*, respectively, and although they do not transmit much water, they influence the movement of water through aquifers. Very few natural materials are completely uniform and most contain aquiclude and aquitard materials.

Figure 1.2 shows how the presence of an aquiclude, such as clay, can give rise to springs and may support a perched water table above the main water table in an aquifer. The top diagram (Figure 1.2a) shows a lower confined aquifer and an upper water table aquifer. The upper aquifer includes low-permeability material that supports a perched water table. The diagram shows the rest-water levels in various wells in both aquifers. Figure 1.2b shows how both confined and unconfined conditions can occur in the same aquifer. In zone A, the aquifer is fully confined by the overlying clay and is fully saturated. The groundwater in this part of the aquifer is at a pressure controlled by the level of water at point p, and water in wells would rise to this level above the top of the aquifer. In zone B, the overlying clay will prevent any direct recharge, although it is unconfined, like zone C. The aquifer in zone C is unconfined and receives direct recharge. Seasonal fluctuations in the water table levels will alter the lateral extent of zone B along the edge of the aquifer. It is likely to be at a minimum at the end of the winter and at its greatest extent in the autumn, before winter recharge causes groundwater levels to rise.

Where impermeable rocks overlie an aquifer, the pressure of the groundwater body can be such that the level of water in wells would rise above the base of the overlying rock (i.e. the top of the aquifer). In such instances the aquifer is said to be *confined*. Sometimes this pressure may be

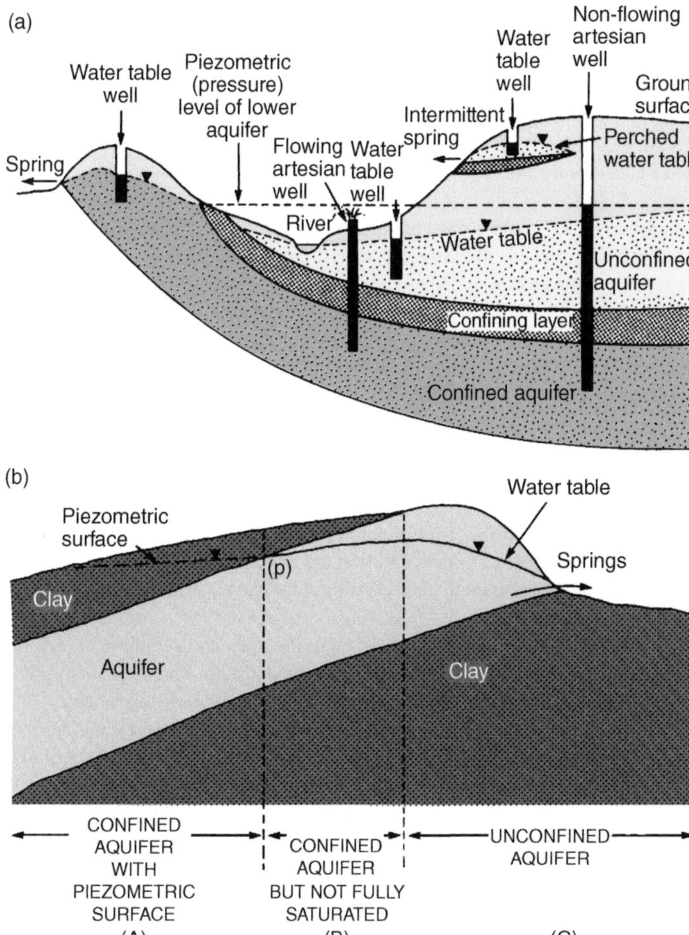

Figure 1.2 *(a) A lower confined aquifer and an upper water table aquifer that includes low-permeability material supporting a perched water table. (b) Both confined and unconfined conditions can occur in the same aquifer.*

sufficiently great that the water will rise above the ground surface and flow from wells and boreholes without pumping. This condition is termed *artesian flow*, and both the aquifer and the wells that tap it are said to be *artesian*.

A groundwater system, therefore, consists of rainfall recharge percolating into the ground down to the water table, and then flowing through rocks of varying permeabilities towards natural discharge points. The flow rates and volumes of water flowing through the system depend upon the rainfall, evaporation, the geological conditions that determine permeability, and many other factors. It is this system that a hydrogeologist is trying to understand by carrying out field measurements and interpreting the data in terms of the geology. The four key factors in achieving a successful investigation are to understand the geology; to interpret the groundwater-level data in terms of the three-dimensional (3D) distribution of heads that drive all groundwater flow systems; to remember that groundwater and surface water systems are interdependent; and to use a structured iterative approach to developing your understanding of the groundwater system you are investigating.

1.2 Conceptual Model

The foundation of all hydrogeological investigations is to gather sufficient reliable information to develop an understanding of how a particular groundwater system works. Such an understanding is usually called a *conceptual model* and comprises a quantified description encompassing all aspects of the local hydrogeology. Consequently, it is necessary for you to think about the way you will develop a conceptual model at the beginning of each project and as the basis of planning the work that is needed.

Although inexorably linked, the activities that form a hydrogeological investigation and the methodology of developing a conceptual model are not exactly the same. The actions at each step of a hydrogeological investigation are generally focused on collecting information, whereas the emphasis in developing a conceptual model is the interpretation of data as they are collected to identify additional information needed to complete the conceptual understanding. A typical hydrogeological investigation can be divided into a number of separate parts, each building on the previous one to eventually achieve an adequate understanding of the system being studied. It will always be necessary to tailor the details of an investigation to the needs of each particular study, although the majority of investigations are made up of the following phases:

- Desk study – the existing available information is assembled to provide an early opportunity to get a 'feel' for the groundwater system and start the conceptual modelling process.
- Walkover survey – it is important to get to know the study area at first hand, so that you can plan your fieldwork programme.
- Exploration – may include drilling boreholes, pumping tests, and geophysical investigations.
- Monitoring programme – defines the variation in groundwater levels, groundwater chemistry, rainfall, spring and stream flows, and so on, both across the area and seasonally.
- Data management – a systematic way of noting data in the field and examining it as it is collected to determine its reliability and if it represents the groundwater in your study area. It is likely that you will store the field data electronically, so do not forget to make regular back-ups.
- Water balance – quantifies the volumes of water that are passing through the groundwater system. Computer simulations may be used in this process to help define recharge and flows through the aquifer.
- Completion of the conceptual model and providing a quantified description of the groundwater system. The quantified aspect is important as it defines things like well yields, groundwater flow rates, recharge quantities and the groundwater chemistry.

A framework for developing a conceptual model as a series of steps was proposed by Brassington and Younger (2010) and is illustrated in Figure 1.3. The steps follow the logical sequence taken in developing a conceptual model that defines the information sources, activities, and review process required along the way, including an audit trail to record all the information that relates to your project. The repetition in the review process is an essential element that ensures the data collection and field work programme are sufficient to enable you to draw meaningful conclusions from your work. The steps summarised in Figure 1.3 are explained in more detail below, with details of how to collect and interpret the data given in subsequent chapters of this book.

1.2.1 Step 1 – defining the objectives

You should define the purpose of each investigation before starting it, so that you focus on all the key questions that need answering as you design the field investigations to provide all the necessary

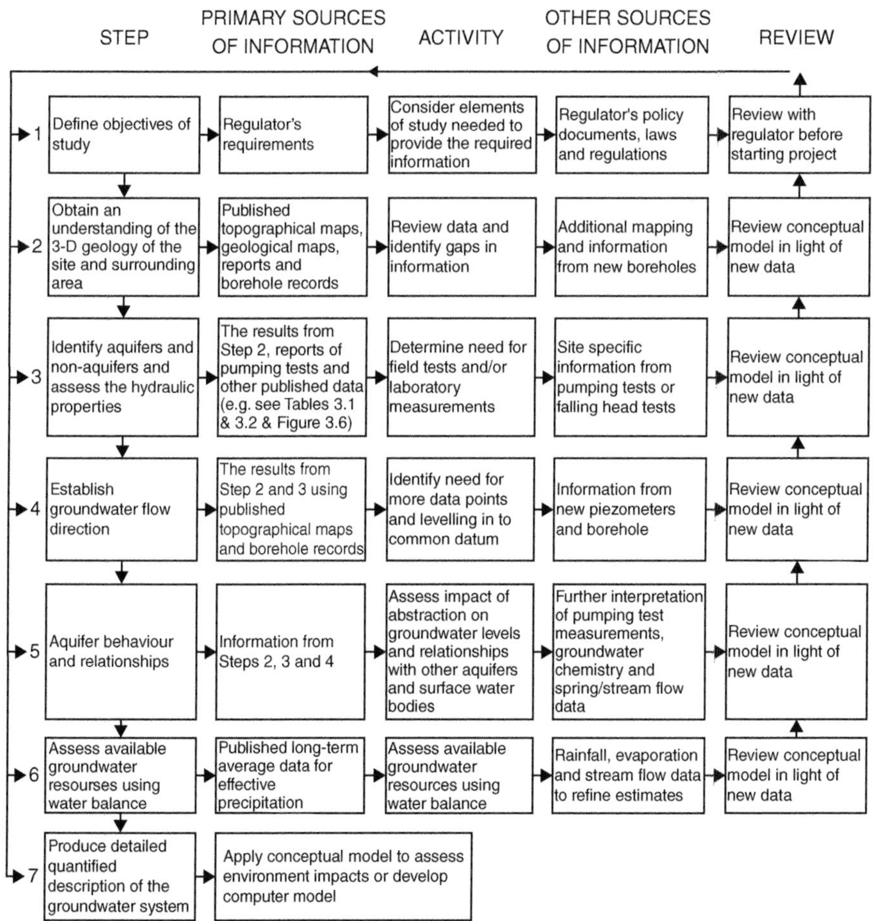

__Figure 1.3__ How a conceptual model is developed from existing information and then gradually improved as field evidence is collected. Figure adapted from Brassington and Younger (2010) with permission of CIWEM.

data. In most cases the objectives should be set out in writing and agreed with your boss, the client, the regulator, or any other people with interests in the outcomes of the project. If you work as a consultant, part of this objective setting may be contained in the proposal that you sent to the client before you were appointed to undertake the project.

1.2.2 Step 2 – defining the geology

The geology of an area controls its hydrogeology and so it is essential to understand the types of rocks present in the area of interest, their lithologies, and their structural relationships. The information can be derived from existing geological maps and reports, although sometimes more information may be needed, possibly involving additional field mapping, drilling exploratory boreholes, or geophysical surveys. Such additional field work should be planned using the results of the desk study, with each element of the new work testing specific aspects of the developing conceptual model.

1.2.3 Step 3 – defining the aquifer framework and boundaries

The aquifer or aquifers being studied exist as 3D bodies and consequently the aquifer boundaries need to be defined on the top, bottom, and all sides. This information is often most easily understood as a series of maps and cross-sections. The geological information derived from Step 2 should be used to identify the aquifers and to estimate the possible values for the aquifer properties, as discussed in Chapter 3. Pumping tests are often required to provide field data from which the hydraulic properties can be calculated. Pumping test methods are described in Chapter 6, along with methods for analysing the data. Copies of the field results and details of these calculations should be kept as part of the audit trail. It is likely that the desk study phase of your investigation that is described in Chapter 2 will include the initial appraisal of the data used in Steps 1–3 of the conceptual modelling process.

1.2.4 Step 4 – defining groundwater flow directions

Groundwater flow directions are best defined using rest groundwater levels measured in non-pumping boreholes that are interpreted using the geological information and information on the aquifers taken from Step 3. Information taken from topographical maps on the location and elevation of springs and surface watercourses is also used, as described in more detail in Chapter 4.

1.2.5 Step 5 – defining the aquifer relationships

This step follows naturally from Steps 3 and 4 and involves considering the flow rates and volumes of groundwater flowing through the system from one part of an aquifer to another, between aquifers and between the groundwater and the surface water systems. Such flow should be quantified and will usually involve calculations based on the Darcy equation (see Chapters 3 and 4). This is the point to decide on the need for a numerical model to assist in the development of the conceptual understanding and/or to make predictions on aspects of the groundwater system when part of it is stressed, such as by new or increased abstraction.

Groundwater systems are usually closely linked to surface water catchments and in order to understand the hydrogeology of an area you should consider both the surface and groundwater catchment areas. Key surface water features such as streams and rivers, springs and ponds, and wetland areas should be identified, initially using topographical maps and later from site visits. The contour information on topographical maps will allow you to define the surface water catchment areas.

New data are likely to be gathered during this step which you should compare with your existing data sets by repeating earlier steps of the framework. The new data may confirm your existing ideas, expand on your developing concepts or even challenge them. This is also the point in the process where you should critically evaluate your understanding of the system to identify any gaps in the available evidence which could be addressed by field investigations. As financial budgets are always important, cost–benefit analysis needs to be taken into account before any decisions are made on this additional work.

1.2.6 Step 6 – water balance

A water balance involves the calculation of the volume of water both entering and leaving the aquifer system being studied, and it is an important factor in all groundwater assessments as it defines the resources available to support abstractions, maintain river flows and wetlands, and provide dilution factors in contamination studies. Where numerical models are being developed, a water balance may be used to gauge the accuracy of the model in replicating the hydrogeological processes involved. The process of undertaking a water balance is described in Chapter 8.

1.2.7 Step 7 – describing the conceptual model

The repeated reviews of the available information and the subsequent collection of the data identified to fill any gaps will eventually result in the development of a conceptual model that is adequate for the purposes set out in the objectives (Step 1). When that point is reached the conceptual model should be set out in your report of the investigation as described in Section 1.4, which will include your answers to the questions originally posed in the objectives. In small-scale studies the description of the conceptual model is likely to be brief and may be simply presented as the conclusions of the report. In larger projects it is likely to be a separate section of the report.

1.2.8 Audit trail

An essential feature of developing all conceptual models, ranging from simple desk studies to complex regional studies, is that they should be auditable. This is not at all sinister! Simply put, it means that someone else can pick up your report and the data sets you have collected and use them to understand how you have arrived at your conclusions. The audit trail merely consists of copies of the correspondence relating to the project, a list of all the information sources you have used, and the records of field measurements. These records include, for example, downloads from data loggers and the certificates of analysis for all the samples you sent to the laboratory. The report for a small-scale study should contain this information as a list of reports and maps used, including the key assumptions made and their justifications, perhaps summarised in a table if there are a lot. Larger studies may involve meetings to discuss progress, at which aspects of the hydrogeological interpretation may be discussed and agreed, with notes kept as part of the audit trail. Very large projects may include small associated studies to review particular aspects of the hydrogeological system, with the reports on such studies forming part of the audit trail. Where a numerical model is developed it is standard practice to keep a record of all model runs, including the aspects of the system being tested by that run and the conclusions drawn from the exercise. These records should form part of the project reports, although they are often kept as reference documents and are not part of either the main report or its appendices.

1.2.9 Quality control

It is important to ensure that the field measurements are made in such a way that the data are as reliable as possible and also that making the measurement has had the minimum possible effect on the data. This means following standard protocols, and is usually best if these are written down and ideally part of widely accepted standards. This approach of formal quality management is used by individual organisations in most countries and includes specifications for performing key activities and for making and maintaining records. In hydrogeological work these may include such tasks as taking groundwater samples, recording data during a pumping test, checking that field equipment is working properly before (and sometimes, after) you have used it in the field, examining data sets to decide that they reflect the conditions in the aquifer, and getting someone to check your report. Of course, having a written method describing how these activities should be carried out does not guarantee that they will be done properly. Neither does it mean that these checks were not a part of routine hydrogeological work before the modern concept of quality control was introduced.

Defining a comprehensive quality control system is beyond the scope of this book. However, the basic checks on equipment are discussed in the appropriate chapters, as well as methods for checking that field readings are taken and recorded properly. If your employer has a formal system, you must ensure that you follow it. Where no formal system is in place, ensure that your personal professional standards include taking proper care in your work and that you ask others to review or audit your work when appropriate. For example, have a colleague read your reports before they are sent out. Use appropriate national standards or codes of practice that are published by bodies such as the British Standards Institution in the UK or the American National Standards Institute in the USA.

There are a number of ways that simply taking field measurements can affect the value you obtain. How to take specific measurements such as a groundwater level in a borehole or piezometer or to collect a groundwater sample are discussed later in this book, where each activity is described. However, there are a few general points that are worth bearing in mind both as you plan your fieldwork programme and when you are out taking the measurements. In hydrogeology a large proportion of field data is collected from boreholes and it is possible that just the presence of the borehole itself changes the groundwater system you want to study. The borehole may connect different aquifers, thereby allowing flow between them along the borehole, potentially affecting both the groundwater level in the borehole and the groundwater chemistry. The direction of flow will depend on the relative heads in each aquifer penetrated by the borehole. Less obviously, a borehole will also provide a flow path when it only penetrates a single aquifer, with the flow driven by the vertical component of the 3D distribution of heads that drive groundwater flow. In recharge areas the flow will be downwards and in discharge areas such as river valleys the flow will be upwards. The reasons for this phenomenon are discussed in more detail in Section 4.7.

1.3 Groundwater Computer Modelling

Computer-based mathematical models are used to make predictions of impacts caused by groundwater abstraction or the movement of pollutants, and may be seen as the end point of a groundwater project. However, the most valuable application of such modelling to groundwater studies is to test aspects of the conceptual model as it is being developed.

Groundwater mathematical models are computer programs based on the groundwater flow equations and a water balance within the aquifer. The aquifer is divided into a large number of segments or nodes that are taken to be representative of a small local area. Values for aquifer properties are assigned to each node, together with values of the recharge from rainfall and other sources and the potential for outflows as abstraction from wells or discharges into the surface water system. The nodes may be of equal size or defined so that some areas of the model are examined in more detail. The water balance equations are solved for each node and the movement of groundwater from each node is calculated. Such flows affect the groundwater levels in neighbouring nodes, requiring the process to be iterative to take account of the water movements through the aquifer. The output is usually in terms of changes in groundwater levels and a statement of the water balance components.

Groundwater models are calibrated using field data. The model-predicted groundwater levels and discharges to surface water are compared with the available field records. Where there is poor correlation between predicted and measured data, the parameters in the model are adjusted and new model runs made repeatedly and compared with the field data until a close fit is obtained.

A number of standard programs (or modelling packages) are commercially available. MODFLOW was originally developed by the United States Geological Survey in 1984 (McDonald and Harbaugh, 1988). It is a 3D finite-difference groundwater model originally conceived solely to simulate groundwater flow. It has a modular structure that has allowed many additional features to be added to simulate coupled groundwater/surface-water systems, solute transport, variable-density and unsaturated-zone flow, aquifer-system compaction and land subsidence, parameter estimation, and groundwater management.

MODFLOW is probably the most commonly used groundwater modelling software in the world. Having had a very large number of users means that most of the errors (bugs) in the program have now been discovered. Versions that run on your laptop are available from several companies, each having front- and back-end packages that format the input data and outputs to suit your needs. It is common for these models to be used at an early stage in an investigation, even on site, as a tool for evaluating field data as they are collected and to assist in the development of the conceptual understanding, as described above. This aspect of modelling may be the most important.

Standard hydrogeological software is also available for pumping test analysis, such as AquiferWin32, for examining water quality data (AquaChem) and to help make impact assessments such as LandSim (used to assess landfill sites) and ConSim (used in contaminated land work). Environmental regulators generally prefer the use of standard software, as it means that the predictions based on the output have the same reliability. Several have assessed the different types of commercially available software and recommend preferred systems. I always try to use the same system as the regulator to minimise disagreements over the interpretation of the results.

Finally, do not be seduced into the trap of thinking that computer modelling is what hydrogeology is really about. The models are only tools to aid understanding and should never be thought of as more than that. High quality computer graphics may look convincing and the models may involve complex and impressive mathematics, but their real value depends on the level of understanding that the hydrogeologist has in terms of the rocks and their influence on groundwater flow. Such an understanding must be based on sound and adequate fieldwork. Do not let anyone ever convince you otherwise!

1.4 Hydrogeological Report Writing

A description of your investigation, the conclusions that you have drawn and the recommendations that follow from them is as important as any other part of the work. After all, a report is likely to be the only tangible thing that is produced by your efforts. Most hydrogeological studies are intended to answer specific questions such as ‘Will a new well on my property produce enough water for my needs?’ or ‘What environmental impact will my new quarry have on water resources?’ and so the report should be written to answer the relevant question. It should have a logical structure and be consistent with the objectives of the study. It should be written in a clear, straightforward way that the reader will be able to understand.

All reports should consist of three parts, namely a beginning, a middle, and an end: in other words, an introduction, setting out the objectives of the study; the main body of the report, which may extend to several chapters; and the conclusions and recommendations. Make sure that you always include these three elements. Table 1.1 sets out a possible structure for a report that provides a logical sequence for you to develop the description of the groundwater system you have been studying.

1.5 Expert Witness

Sometimes a hydrogeologist is asked to be an expert witness and to give evidence to the court or a public inquiry, and to understand what is required we first need to sort out a few definitions. An *expert* is someone with experience in a particular field or discipline beyond that expected of a

Table 1.1 *Template for planning the contents of a report.*

Section	Content
Summary	Must be written last when the rest of the report is complete and cover the main points contained in the body of the report. It is important as many recipients of your report will only read the summary. Do not include material or ideas that do not appear in the body of the report.
Introduction	Explains the objectives for the study and provides the basic logic to the report structure.
Topography and drainage	A description of the topography and surface water systems will help to define the aquifer boundaries and quantify the available water resources.
Geology	The description of the geology should be in sufficient detail to identify the main aquifers and factors influencing the groundwater system. Maps and cross-sections are very good ways of explaining the geology and help non-geologists to understand your arguments.
Hydrogeology	Describe the aquifers and their hydraulic properties and whether the flow is intergranular or through fractures. Define flow directions and discharge points.
Groundwater chemistry	Use the chemistry to characterise different aquifers. Piper, Schoeller and other graphical representations will help explain your arguments.
Water resources	Define the available resources and how much is needed both to support wetlands and surface water flows and to supply existing abstractors.
Risk assessment	Use a systematic approach to quantify the risk of a new activity having a serious impact. Vulnerable features are identified and the potential risk considered for each one. Pollution studies use the Source–Pathway–Receptor model.
Conclusions and recommendations	Conclusions should be based on and supported by the information described in the main part of the report. The recommendations should match the objectives set out in the introduction.
Appendices	Appendices are best used for those parts of your work that do not easily fit into the main text, such as calculations or data lists.
Tables and diagrams	A picture is worth a thousand words and so are tables. Both provide excellent ways of summarising information and presenting your arguments.

layman. An *expert witness* is an expert who makes this knowledge and experience available to a court or other judicial or quasi-judicial bodies, such as a parliamentary committee or a public inquiry, to help it understand the issues in a case, so that it can reach a sound and just decision. An expert witness is paid for the time it takes to form an opinion and, where necessary, support that opinion during the course of litigation.

An expert witness is not paid either for the opinion given or for the assistance that opinion gives to the client's case – that is different to being paid for the time taken to form your opinion. It is very important to keep these definitions clear because if you stray from acting as an expert witness into advising the client you become an *expert advisor* and your immunity from prosecution will be removed. You must remain aware of the distinction at all times, and only move into the role of expert advisor in full knowledge of the legal consequences.

An expert might take on a number of roles. An expert instructed by just one party in a claim, and whose opinion is to be put before the court, is a proper expert witness. When advising a party, without any intention of putting the expert's opinions before the court, the expert is known as an expert advisor or shadow expert. In this case he or she may attend the court to advise the lawyers who ask the questions during cross-examination but will not speak to the court. If the expert witness is instructed by all the parties in a claim, then he/she acts as a Single Joint Expert. This latter case is preferred by the courts particularly in small cases as it reduces the costs and also focuses on the main issues more easily.

The fundamental characteristic of expert evidence is that it is opinion evidence. Generally speaking, lay witnesses may give only one form of evidence, namely evidence of fact. They may not say, for example, that a vehicle was being driven recklessly, only that it ended up in the ditch. In this example, it is the task of the expert witness to assist the court in reaching its decision with technical analysis and opinion inferred from factual evidence of, for example, skid marks.

To be of practical assistance to a court, however, expert evidence must also provide as much detail as is necessary to allow the court to determine whether the expert's opinions are well founded. It will often include factual evidence supplied in the expert's instructions which requires expertise in its interpretation and presentation; other factual evidence which, while it may not require expertise for its comprehension, is linked inextricably to evidence that does; explanations of technical terms or topics; and hearsay evidence of a specialist nature, such as the consensus of medical opinion on the causation of particular symptoms or conditions, as well as opinions based on facts put forward in the case.

Expert evidence is most obviously needed when the evaluation of the issues require technical or scientific knowledge only an expert in the field is likely to possess. However, there is nothing to prevent reports for court use being commissioned on any factual matter, technical or otherwise, providing that it is deemed likely to be outside the knowledge and experience of those trying the case, and the court agrees to the evidence being called.

In all cases, you will be expected to prepare a report and send it to the court or to provide it as evidence on which you will be examined at a public inquiry. It is important to make your report as simple as possible so that it is easily understood by all those who read it, which is often done by defining all the technical terms that you use the first time you use them. It is very important to take your time in preparing the report and to get someone else to read it for you. If you work in a consultancy practice you may well have a quality control system that takes care of such things. However, the people who read the report will also use the same language as you and so it may be advisable to have additional people read it that do not have the technical language to have that extra check on clarity. What seems like everyday language to you may well be difficult to understand by a layperson.

You may not always be called to give evidence, for several reasons. The parties may be negotiating over a claim, say, and the report may be used as a bargaining tool. In other instances, although your report has been accepted by the court, there are no aspects of it that require further discussion. This has happened to me where I showed in a report that several water samples had the same chemistry using Piper and Schoeller diagrams and that was all the court needed to make a decision without me being called to give evidence.

If you do attend court or a public inquiry, make sure that you wear a suit (and a tie if you are a man) and make sure that you are polite and do not allow yourself to get emotional. Giving evidence is under an adversarial system, which can be off-putting to most people, especially the first time that you do it. The lawyers for the 'other side' may ask you questions on what in normal circumstances you may regard as offensive; do not be put off, that is what they are trying to do. Not all situations are like that. For example, in some situations where there are not many people likely to want to attend, a hearing may be held where an inspector and the representatives of all sides sit round a table. In most cases, however, an inquiry is likely to be held in a larger hall near the site in question, with each side taking up its own space. Court hearings are always heard in a court.

One of the most important things you need to do is to visit the site just before you give evidence so that you can answer questions on its present condition. I have heard of experts not doing that and suffering the embarrassment of looking foolish when cross-examined. Seeing the site will avoid that problem, will bring it all to mind and will help you to keep calm and collected!

2
DESK STUDY

A desk study is the preliminary examination of the hydrogeology using all the available information before additional data are obtained from the field. It is usually carried out in an office, hence the name. The desk study is the first step in developing your conceptual model, and the methods described in this chapter are likely to be the first steps in any hydrogeological assessment. By defining what is already known about an area and identifying which questions need answering, a desk study provides a very useful way of planning a fieldwork programme. An early appreciation of the different field measurements needed will enable equipment to be assembled and plans to be made, so that the fieldwork can be completed over the minimum time period possible. Some measurements can only be made during a particular season or to fit in with other activities, and planning will avoid missing opportunities and delaying the completion of your work.

You will need information on a wide variety of topics, including geological maps and reports, topographical maps, borehole records, rainfall and river flows. The availability of this information will be very variable. Some useful data may have been collected by government departments, local government organisations and a variety of other bodies for a large number of different purposes, some of which may have little directly to do with groundwater studies. In most cases you will have to supplement existing information with your own field measurements.

If you are unable to obtain information from organisations within a particular country, you may be able to obtain it from American (USA) or European Union (EU) organisations. The geological survey and university departments from these countries have undertaken overseas work for many years. Besides providing maps and reports, these organisations may help you to locate a source of information in the country where you are working. Often the easiest and fastest way of finding a data source or even the data itself is to search the Internet using one of the commonly available search engines. There are a large number of on-line bibliographic databases that can help you look for references and companies that will conduct literature searches on specified topics. Your local librarian is likely to have information on these services. For example, working in the UK, I use Streetmap UK (www.streetmap.co.uk) to look at the area on a 1:25,000 scale Ordnance Survey map and the British Geological Survey site (https://www.bgs.ac.uk/bgs-geology-viewer/ and https://mapapps2.bgs.ac.uk/geoindex/home.html?layer=BGSBoreholes&_ga=2.18562810.2139103869.1688208829-1913099086.1688208829) is useful in taking a first look at the geology and also for borehole records.

Use a systematic approach to the desk study and methodically examining all the available information. Where little information is available on an area, the desk study will be completed quickly and only produce a simple idea of how the groundwater system works. The desk study should be completed nevertheless, as it plays an essential part in planning what may be an extensive field investigation.

2.1 Defining the Area

Topographical and geological maps are the basic information source to define the extent of the area you need to include in your study. In the UK the Ordnance Survey 1:25,000 scale maps are usually the best to use as they show field boundaries and also most water features. The size of the area

Field Hydrogeology, Fifth Edition. Rick Brassington.

depends on the reason for the project. If you are assessing possible impacts of an abstraction of less than 20 $m^3 d^{-1}$ from a new borehole, an area with a radius of 100 m may be adequate, except in some fissured formations, where a larger area may be needed. If the work is being carried out as part of a waste disposal investigation, extend the study area well beyond the immediate area of the site. Never regard the boundaries as rigid, and where wells or springs lie just outside the periphery of the study area, for example, it is usually best to include them.

When assessing the overall groundwater resources of an aquifer you should include the whole aquifer formation and extend it on to neighbouring strata to include any cross-boundary flows. Attention must be paid to stream catchment boundaries, which are fixed by the topography and may not coincide with groundwater or aquifer boundaries. Such investigations may extend over an area less than 1 km^2 in the case of a minor aquifer such as a river terrace, to several thousand square kilometres for major aquifers.

2.2 Identifying the Aquifers

One of the first steps in a hydrogeological study is to identify the aquifers and aquitards. Use all the available geological information to classify the rocks in the area as good aquifers, poor aquifers or non-aquifers, even preparing a simplified version of the geological map using these categories. Next, consider the aquifer boundaries. There are many sorts of boundary that in broad terms can be classified as permeable or impermeable.

Figure 2.1 shows a few examples of each type of aquifer boundary, but remember that there are many more. The boundary between the two aquifers faulted against each other (a) may be permeable and allow flow across it, with permeability being locally increased if rock fracturing has occurred. Alternatively, the boundary may be impermeable due to secondary minerals deposited along the fault plane or the presence of a fault gouge. An impermeable boundary may be detected by differences in groundwater chemistry or from pumping test analysis. The faulted boundary shown in (b) is essentially impermeable, as the aquifer is faulted against an aquiclude. In example (c), aquifer A lies unconformably over a sequence that includes two separate aquifers, B and C. Conditions are similar to example (a), except that groundwater quality may be different in all three aquifers. In example (d), the groundwater conditions in the aquifer are similar to those in (b) in that the aquifer unconformably overlies a non-aquifer. In example (e), an unconsolidated aquifer, such as a river terrace, partly overlies a lithified aquifer (such as limestone or sandstone) that forms an escarpment. Groundwater flow is possible between the two aquifers and, because of the topographic differences, it is most likely to be from aquifer B to aquifer A. In these circumstances, scarp-slope springs may not be seen as they may be buried beneath the river terrace. In example (f), a gravel aquifer overlies low-permeability bedrock. No groundwater flow takes place across the boundary, but the aquifer recharge will be greatly enhanced by surface runoff from the upland area.

Decide whether it is possible for groundwater to flow across the aquifer boundary, as this will affect groundwater recharge and flow directions, and there are also implications for groundwater quality and the pumping characteristics of wells.

Next look at the thickness of the aquifer and how it varies across the area, as this has a significant effect on the transmissivity. Borehole information will be particularly useful in this exercise and helps define the base of the aquifer. Some formations, such as mudstones, igneous and metamorphic rocks, may act as minor aquifers only when they are fractured or weathered. Borehole logs may provide information on the depth of weathering. In high latitudes, perennially frozen ground (termed *permafrost*) has a significant effect on aquifer characteristics and groundwater conditions. Frozen ground has a lower permeability compared with the unfrozen material, with the differences often being several orders of magnitude (factors of tenfold). Consequently, permafrost acts as an aquiclude or an aquitard, reducing recharge and confining the groundwater in the deeper unfrozen aquifer.

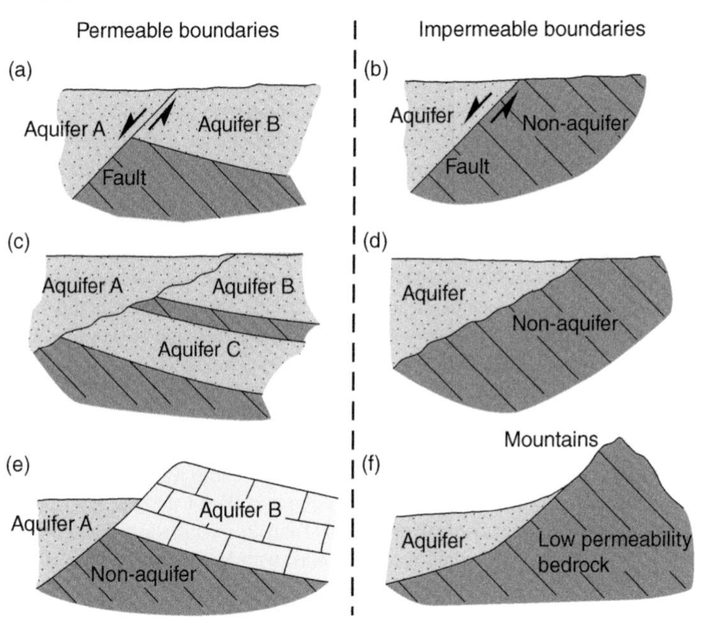

Figure 2.1 *Different types of aquifer boundaries are shown in this diagram and discussed in the text.*

Do not neglect the upper boundary of the aquifer. Is it at outcrop or is it overlain by other deposits? Outcrop areas are potential recharge areas, but any overlying low-permeability rocks may confine the aquifer, thereby having a fundamental influence on groundwater conditions. Bear in mind that the unsaturated part of an aquifer formation will become saturated as groundwater levels rise in response to seasonal recharge or a reduction in groundwater abstraction. Similarly, the upper part of the saturated aquifer may become dewatered in response to seasonal fluctuations or increased abstraction.

At this stage you should consider the need for a pumping test or tests on new or existing wells (see Chapter 6). The information derived from such tests can help define the aquifer boundary conditions, enable you to estimate the hydraulic properties of the aquifer and provide straightforward information on the hydraulic relationships within the groundwater and surface water systems.

2.3 Groundwater Levels

At the start of an investigation it is unlikely that many direct groundwater-level readings are available. However, information from the topographic map may provide a general idea of groundwater discharge points and flow directions. Spring lines may be identified using the 'springs' marked on the map, together with the starting points of streams that can be taken as the end of the line (often blue) that represents a river. Poorly drained areas may be shown as 'marsh' or identified by a concentration of ditches (shown as straight blue lines on the map). They may be underlain by low-permeability materials and may mean that groundwater is discharging from an adjacent aquifer. Comparison of such poorly drained areas with the geological map will help distinguish between aquifer discharge areas and boggy patches caused by underlying low-permeability materials.

Compare maps published on different dates to see whether the areas of poor drainage have changed over the years. A reduction in the size of marshes could indicate improved drainage and may be associated with an increased number of ditches shown on the map. Alternatively, shrinking marsh areas may be a result of a lowered water table caused by high rates of groundwater abstraction.

Groundwater flows from areas of recharge to areas of discharge, with the general shape of the water table usually being a subdued version of the surface topography. Hence, once discharge areas have been identified as springs, streams or even along the coast, it is possible to infer the general direction of groundwater flow from the topography. This exercise will provide a clue as to where the recharge areas lie in the upstream direction of the groundwater flow. Recharge areas may be also indicated by a general lack of streams and other surface water features. Significant natural recharge can only occur at outcrop or through permeable superficial deposits; therefore, a further comparison of the geological and topographic maps will enable possible recharge areas to be identified and groundwater flow directions to be deduced. Beware of having too much confidence in your deductions at this stage! Treat them as a hypothesis that must be tested using field evidence. Be particularly careful in fissured aquifers, especially karstic limestones, as groundwater can flow in different directions from those indicated by signs at the surface.

Figure 2.2 shows an example of how groundwater flow directions were deduced at the desk study stage of an investigation of the St Bees Sandstone aquifer in West Cumbria, England. The aquifer lies unconformably on older rocks (mainly mudstones) and dips to the west. The aquifer outcrops in areas of higher land that form the main recharge areas, with local springs, streams and the coast providing potential discharge areas (see also Figure 4.27).

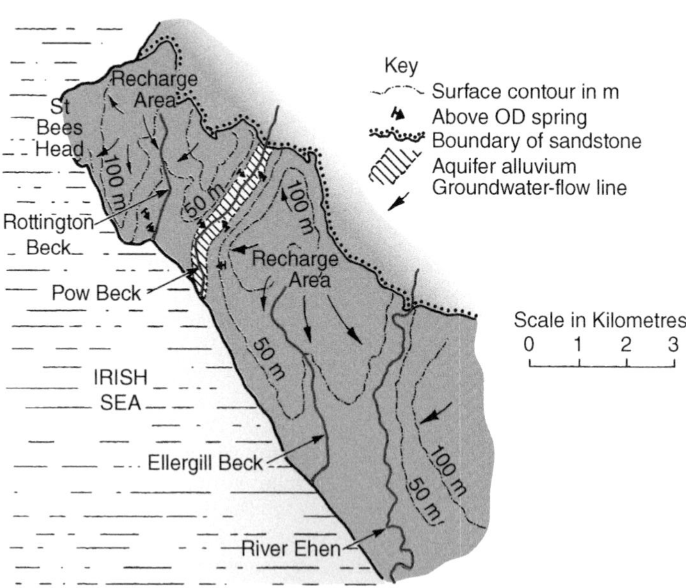

Figure 2.2 *This groundwater flow map for part of the Permian St. Bees Sandstone aquifer in West Cumbria, England, was drawn at the beginning of the desk study stage to provide an early idea of groundwater levels before any boreholes had been drilled. Also see Figure 4.27. (Data by courtesy of the Environment Agency.)*

2.4 Surface Water

In many areas, rivers and lakes receive an important component of flow from groundwater, which allows groundwater discharges to be quantified from surface water flow data. Those streams and rivers that drain water from aquifers in the study area should be identified, and any flow records examined using the method outlined in Section 5.7. Once the available information has been compiled, the need for additional flow measurements can be assessed. Sometimes it will be sufficient to measure the flow upstream and downstream of the study area to see whether flows increase. Where large-scale pumping tests on new boreholes are involved, continuous records may be required to see if the new pumping reduces stream flows. Such measurements are usually inconclusive, however, unless the pumping rate is equivalent to at least 50% of the dry-weather flow (i.e. the lowest natural flows) of local streams and the test continues over a few weeks.

2.5 Recharge

Information is needed on rainfall and evaporation to estimate aquifer recharge. Assess whether or not existing rain gauges and the nearest meteorological station where evaporation is measured are close enough to the study area to provide accurate information, and consider adding extra ones more locally.

To decide whether there are enough rain gauges, compare both the annual average values for all rain gauges in the vicinity and their altitude. Areas of higher altitude generally have a higher rainfall. Increased precipitation with altitude can be significant in areas with large variations in topography, as there may be more than twice the rainfall on top of a hill compared to its base. Where a groundwater investigation is likely to go on for any significant period, it is often worthwhile setting up at least one new rainfall station to supplement sparse data.

Evaporation measurements are difficult and it is best to use information from an established meteorological station if you can. Remember that these values represent the potential evaporation, that is, the maximum that can occur. Actual values of evaporation might be much less, as they are limited by the quantity of the water available. Evaporation will stop when all the water has gone!

It is possible to make an initial calculation of annual recharge using average values for rainfall and evaporation. The estimate will be somewhat crude, but an early idea of recharge values will be useful where a new groundwater development is proposed. If the estimated recharge is several times larger than the quantity of water required for supply, then the scheme is likely to succeed, and, conversely, schemes that need all the available recharge are unlikely to be successful.

To make the calculation, subtract the evaporation value from the average annual rainfall to obtain an idea of how much rainfall is available for recharge (i.e. the potential recharge). Then multiply the potential recharge value (expressed as a depth of water in millimetres) by the area of the exposed aquifer. Although it is assumed that all the potential recharge will percolate into the outcrop areas, a smaller amount will soak into the aquifer where it is overlain by low-permeability material. It should be supposed that no recharge will take place through indurated mudrocks or where the aquifer is confined by thick clays. If the aquifer is not confined (see Figure 1.2b), assume that half the potential recharge will reach the aquifer by percolating through the clay or by running off the edge of the clay onto the outcrop. Recharge can occur through very low-permeability materials provided that the vertical hydraulic gradient permits it. This means that the groundwater head in the aquifer must be at a lower level than that in the overlying rock.

The quantities may be calculated using Darcy's law. Such recharge is low and may be in a range of 10–50 mm year^{-1} of the available rainfall. However, over large areas these small rates provide significant volumes of recharge water. In urban areas, recharge is reduced to about 25% because rainfall runs off buildings and roads into sewers. An example is given of this type of calculation in Figure 2.3. In this example the annual average quantity of rainfall available as aquifer recharge has been calculated in Steps 1–3. The aquifer is then examined and the area of different types of surface condition are measured and summarised in Step 4. In this example it is assumed that clayey drift

1. Average annual rainfall			987 mm
2. Average annual potential evapotranspiration			450 mm
3. Potential recharge:	987 – 450	=	537 mm
4. Area of aquifer			
(a) Outcrop			15.7 km^2
(b) Overlain by clayey drift			36.3 km^2
(c) Confined by mudstones			not used in calculation
(d) Outcrop with urban areas			2.5 km^2
(e) Clayey drift with urban areas			3.2 km^2
5. Calculation of average annual recharge			
(a) Outcrop	15.7 × 537	=	8431 × 10^3 m^3/year
(b) Clayey drift	36.3 × 537 × 0.5	=	9747 × 10^3 m^3/year
(c) Confined		=	nil
(d) Outcrop with urban areas		=	1007 × 10^3 m^3/year
(e) Clayey drift with urban areas		=	644 × 10^3 m^3/year
	Total	=	19829 × 10^3 m^3/year

Figure 2.3 *Calculation of annual recharge.*

reduces the recharge by 50%. The information is summarised in Step 5 to calculate the annual recharge as approximately 19,800,000 m^3 year^{-1} or an average of 54,000 m^3 d^{-1}.

A useful comprehensive guide to estimating recharge is given in *Groundwater Recharge* (Lerner et al., 1990). In the UK, the Meteorological Office publishes 25-year average values of rainfall, evaporation from grassland and effective rainfall under the acronym MORECS (Meteorological Office Rainfall and Evaporation Calculation System) (Hough & Jones, 1997). These data are available from the Meteorological Office as long-term average figures over MORECS squares that have sides 40 km long, or as site-specific data covering specific time periods and areas. You may have to purchase some or all of these data. Don't forget climate change where rainfall intensity is important.

2.6 Groundwater Use

The availability of groundwater abstraction records is extremely variable. At best you are likely to find that records are only kept for the largest abstractors, and an on-the-ground survey will be necessary to locate the rest so that you can estimate the quantities involved. Information on these larger abstractors will be useful to give an early idea of borehole yields in the area, and will indicate whether there are any existing abstractors who may be affected by a new borehole source or threatened by a new waste disposal site, for example. Estimate the quantities abstracted using the information in Table 2.1.

2.7 Groundwater Chemistry

Except for public water-supply sources, it is unusual for much groundwater chemistry information to be available in existing records. Sometimes information is tucked away in the abstraction records for the borehole, or in a notebook where water levels are logged, or it may be in the environmental health department of the local council. The only way to find them is to painstakingly go through all the record books and files that are available. Old records are unlikely to quote values in milligrams per litre, so see Table 7.13 for conversion factors.

Table 2.1 Average water requirements for various domestic purposes, agricultural needs, and manufacturing processes.

Use of product	Quantity of water needed (L)
Domestic use (per person per day)	
Drinking and cooking	4
Washing and bathing	45
Flushing lavatory	50
Cleaning and washing up	14
Total average daily requirement per person	135
Animals (daily requirements)	
Cow (milk producer including dairy use)	150
Beef cattle	25–45
Calf	15–25
Horse	50
Pig	15–20
Lactating sow	15–30
Weaners (young pigs)	5
Sheep (drinking)	2.5–5
Sheep dipping (per dip)	2.5
Poultry layers (per 100 birds)	20–30
Poultry, fattening (per 100 birds)	13
Turkeys, fattening (per 100 birds)	55–75
Manufacturing	
Paper/tonne	2500
Bricks/tonne	320
Steel/tonne	1600–3300
Cosmetics/tonne	3300
Car (each)	4500
Commercial vehicle (each)	2700

Based on Brassington (1995) and data from Internet sources.

Be on the alert for potential problems such as saline intrusion, high nitrates from agricultural fertilisers, leachate from waste disposal sites, sewage effluent – especially septic tanks and cesspits – runoff from mineral workings tailing dumps and the potential for contamination from pesticides and solvents. All of these could cause serious problems for both new groundwater developments and existing supplies, and additional monitoring may be needed.

2.8 Aerial Photographs and Satellite Imagery

Although largely superseded by satellite imagery and other techniques such as LiDAR (light detection and ranging), aerial photographs remain of potential value in hydrogeological field work. Lisle et al. (2011) give a good description of the uses of remote sensing for field geologists, including both aerial reconnaissance and satellite imagery.

Satellite images are readily available and can be purchased as photographic prints or digital images downloaded from the Internet. All satellite images produced by NASA (the US government agency National Aeronautics and Space Administration) are published by its Earth Observatory and are freely available to the public. Several other countries have satellite imaging programmes, such as the European Space Agency's ERS and Envisat satellites that carry various sensors useful across a range of environmental applications. There are a number of private companies that provide commercial satellite imagery, and in the early twenty-first-century satellite imagery is widely available on-line through systems such as Google Earth, Google Maps (www.google.co.uk/maps) and Microsoft's Bing Maps (www.bing.com/maps). The Google Maps systems include 'Street View', which are images taken from public roads at street level, and Bing Maps provides an alternative to satellite images, with images taken from low-flying aircraft at an oblique angle to give a 3D effect.

LiDAR uses readings taken with instruments in survey aircraft to measure the shape of the ground surface, including natural and man-made features. A survey could be carried out especially for a project, although libraries of survey images are growing and can be purchased from specialist survey companies. The technique measures heights and elevations that are accurate to between 10 and 20 cm, which is sufficient accuracy for identifying surface catchment boundaries and the elevation of springs and borehole wellheads in hilly areas where there are large differences in elevation from one site to another. More accurate values will require surveying techniques (see Section 4.3).

Aerial and satellite images are particularly useful in a desk study, because they may be very detailed and show up features that cannot be seen easily on the ground. The general approach is to use the images to prepare maps showing variations in vegetation type, landforms, land use, soils, and drainage. These maps are then used to interpret likely groundwater conditions, from which it is possible to find seepage areas to define the best areas for new wells, for example.

Both satellite imagery and LiDAR data are in a digital format, allowing data to be processed to give prominence to particular features such as temperature, and are very much more sensitive than the old-fashioned infrared film. The use of false colour helps distinguish differences that may not be seen in any other way. LiDAR and satellite images may cover many square kilometres and can be stored electronically on laptops and taken to site. You can use them to 'zoom out' and see things that are not obvious from the ground and even have a peek over the horizon, all without moving out of your chair! These methods help you plan the next bit of fieldwork and help you make the best use of time in the field. These images will allow you to spot key features, such as sources of contamination in those areas where you are unable to obtain access. There is good coverage in western countries, and you can order up bespoke images for other parts of the world, often where map coverage is poor, and these images are worth their weight in gold!

2.9 Planning a Fieldwork Programme

Having completed your desk study you will have examined all the available information and identified what extra measurements are needed. An early objective of the fieldwork programme must be to obtain a first-hand knowledge of the area. If it is small enough, familiarise yourself with the area by walking over it when carrying out surveys to search for wells and springs. If you have a very large area to investigate, it will be necessary to use a vehicle or perhaps even a helicopter, but try to walk over as much as possible. There is no substitute for personal observation to obtain a detailed knowledge of your area or an understanding of how the groundwater system works. By this direct experience, hydrogeological skills and knowledge are built up that will enable you to interpret geological information in groundwater terms. This is by far the best way to become a competent professional hydrogeologist. A checklist for planning fieldwork programmes is contained in Table 2.2.

Table 2.2 *Checklist for planning a fieldwork programme.*

Topographic information
Are adequate maps available? If not, use aerial photographs as a substitute base-map. Supplement with levelling and other field observations where necessary.

Geological information
Is the available information adequate to define aquifer boundaries? Are either additional geological mapping or boreholes needed to provide geological information?

Groundwater levels
Carry out an on-the-ground survey to locate and record the position of all springs, wells, and boreholes. Do topographic maps or aerial photographs show seepage areas? Can you draw reliable groundwater-level contours with the available data, or are extra boreholes needed? Decide on the need for a monitoring programme and details of frequency of observations and the equipment required.

Surface water measurements
Are extra flow measurements required? Decide on suitable gauging sites and methods, frequency of measurements, and the equipment that you will need.

Rainfall and evaporation
Are there adequate rain gauges in the area? Where is the nearest meteorological station which provides evaporation figures? Do you need to take your own measurements? If you do, then decide on suitable sites for your instruments.

Groundwater use
Collect information on water abstraction quantities during the well-location survey.

Pumping tests
Will pumping tests be needed to provide data on aquifer properties and/or to determine the environmental impact of a new abstraction? Make sure that you use the opportunity provided by the test to obtain groundwater samples for analysis.

Groundwater chemistry
Are extra samples needed? If yes, incorporate sampling into other fieldwork programmes.

3
FIELD EVALUATION OF AQUIFERS

Geological properties such as lithology, petrology, and structure largely control water flow through the ground and influence both the direction of groundwater flow and the yield of wells. The characteristics that define groundwater flow and storage are usually referred to as the *hydraulic properties*. These can be measured in the field or laboratory, but can also be assessed in general terms by consideration of the overall aquifer geology. Field measurements of the hydraulic properties are made by means of pumping tests that are described in Chapter 6.

3.1 Grain Size Analysis

Aquifer permeability can be inferred from its lithology, and this may be used as the basis of an initial assessment. A grain-size chart (see inside front cover) is a very useful tool for estimating particle size in both unconsolidated and indurated sediments by simply comparing a small sample with the chart. The most precise value for the grain-size distribution of a sample is obtained by sieving in a laboratory.

Small sieves that can be used in the field (see Figure 3.1) are better than a simple estimate using the grain-size chart. The photograph (a) shows examples of sieves where the sample is brushed through the sieves to separate that part of the sample in each grain size. The results can be used to produce a particle-size distribution by plotting a graph on semi-logarithmic paper with the particle sizes on the logarithmic scale. The data are expressed as the percentage weight passing each sieve plotted on the other axis (Figure 3.1). The three example curves show a uniform sand (1), a poorly graded fine to medium-coarse sand (2) and a well-graded silty sand and gravel (3). Hydraulic conductivity can be estimated from the grain-size value at 10% (also called the effective grain size or D_{10}). For example, the uniform sand (1) has a D_{10} of approximately 120 μm (0.12 mm).

Alternatively, an improvised method can be used based on the principle that fine grains settle out in water more slowly than coarse grains, and can provide an acceptable alternative when sieving is not possible. Figure 3.2 illustrates the method with an example. You can assess the grain size of a sample using a narrow, parallel-sided glass jar with a screw lid, a grain-size chart (endpaper), a hand lens (×10) and a ruler. Half fill the jar with a representative sample and then fill it to near the top with water, adding a few drops of water glass (sodium silicate, Na_2SiO_3) if available. This will act as a flocculation agent, helping the suspended grains to settle faster. After securing the lid, shake the jar vigorously to ensure that all the material is in suspension and then allow it to settle on a flat base for 24 hours or until the water is clear. The sample should then be graded in a fining-upward sequence. Examine the sample through the hand lens using the grain-size chart to help distinguish between clay/silt, fine/medium sand and coarse sand. Take care when handling the jar to avoid disturbing the sample. Measure the thickness of each layer to estimate the proportion of each grain size in the sample. In the example shown, 30% is clay and silt, 51% is fine/medium sand and 19% is coarse sand. Remember, field tests are only approximate and the results cannot be used with the same confidence as those from laboratory measurements.

Field Hydrogeology, Fifth Edition. Rick Brassington.

(a)

(b)

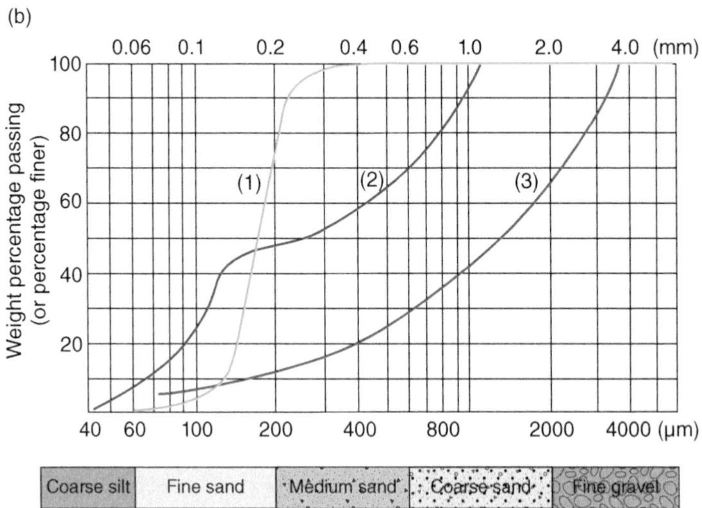

Figure 3.1 *(a) Examples of sieves suitable for testing small samples in the field. (b) The three example curves show a uniform sand (1), poorly graded fine to medium-coarse sand (2) and well-graded silty sand and gravel (3), which are discussed in the text. (Courtesy of Endecotts Limited.)*

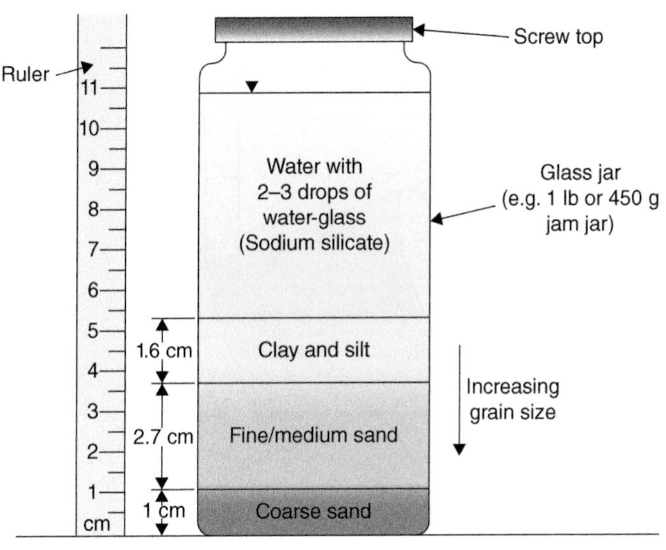

Figure 3.2 *An improvised method to assess grain-size distribution based on the principle that fine grains settle out in water more slowly than coarse grains.*

3.2 Hydraulic Properties of Aquifers

Much of the quantitative study of groundwater flow is based on the results of experiments carried out in 1854 by Henry Darcy, a French hydraulics engineer who worked in a water treatment works at Dijon. He was interested in how different sizes in the filter bed, the grain size of the sand and the head across the filter changed the flow rate through sand-filter beds.

Groundwater flows through an aquifer are driven by the imbalance in water pressure (or *head*) over the aquifer. The difference in groundwater levels is called *head loss* (h) and is usually expressed in metres. The slope of the water table is called the *hydraulic gradient* (h/l), and is the dimensionless ratio of head to distance (Figure 3.3).

The equation that relates the groundwater flow rate (Q) to the cross-sectional area of the aquifer (A) and the hydraulic gradient (h/l) is known as the *Darcy equation* (*or Darcy's law*) and has the following form:

$$-Q = KA\frac{h}{l}$$

The negative sign is for mathematical correctness and indicates that the flow of water is in the direction of decreasing head. From a practical point of view, it may be ignored. In the equation, K is the *hydraulic conductivity*, defined as the volume of water that will flow through a unit cross-sectional area of aquifer in unit time, under a unit hydraulic gradient and at a specified temperature. The usual units of hydraulic conductivity used by hydrogeologists are metres per day ($m\ d^{-1}$) (this unit is a simplification of $m^3\ d^{-1}\ m^{-2}$). Hydraulic conductivity is also expressed in metres per second (i.e. $m\ s^{-1}$, which is a simplified form of $m^3\ s^{-1}\ m^{-2}$), and there are other units that are included in Appendix B.

Darcy's equation can also be written as:

$$v = \frac{Q}{A} = -K\frac{h}{l}$$

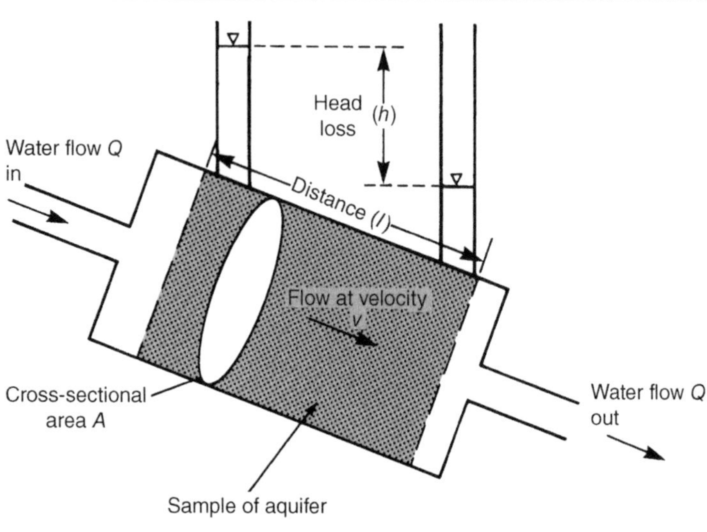

Figure 3.3 *Terms in the Darcy equation. See text for explanation.*

In this equation v is the *apparent velocity* of the water flow, also known as the *Darcy velocity* or *groundwater flux*. The equation assumes that the flow takes place over the whole cross-sectional area of the aquifer and ignores the relative proportion of the solid parts to the pore spaces. In reality, the flow is restricted to the pore spaces, so the actual average velocity (V_α) is much greater than the Darcy velocity and is defined as

$$V_\alpha = \frac{Q}{\alpha A}$$

where α is the effective porosity of the aquifer (expressed as a decimal or a percentage).

Hydraulic conductivity depends on both the properties of the aquifer and the density and viscosity of the water. These properties of water are affected by a number of conditions, such as the concentrations of dissolved minerals and temperature. An increase in water temperature from 5°C to about 30°C, for example, will double the hydraulic conductivity and will double the rate of groundwater flow. Groundwater temperatures remain relatively constant throughout the year in deep aquifers, so these changes are not normally a problem for hydrogeologists. In some shallow aquifers in areas of climate extremes or in particular situations involving waste hot water and industrial effluent, the flow rates may be affected by the temperature. When hydraulic conductivity is being measured by pumping water into a test section or sample either in the laboratory or in the field it is important to ensure that the temperature of the test water is the same as the usual groundwater temperatures in the aquifer being investigated, or apply a correction factor to the results.

The property of a rock that controls the hydraulic conductivity is its *intrinsic permeability* (k), and is constant for an aquifer regardless of the fluids flowing through it, applying equally well to oil, gas and water. Intrinsic permeability can be calculated when fluid density and viscosity are known. It has the reduced dimension of metres squared but is usually expressed in darcys (1 darcy = 0.835 m d^{-1} and 1 m d^{-1} = 1.198 darcy for water at 20°C).

The amount of water held in a rock depends upon its *porosity*. This is the proportion of the volume of rock that consists of pores, and is usually expressed as a decimal or percentage of the total

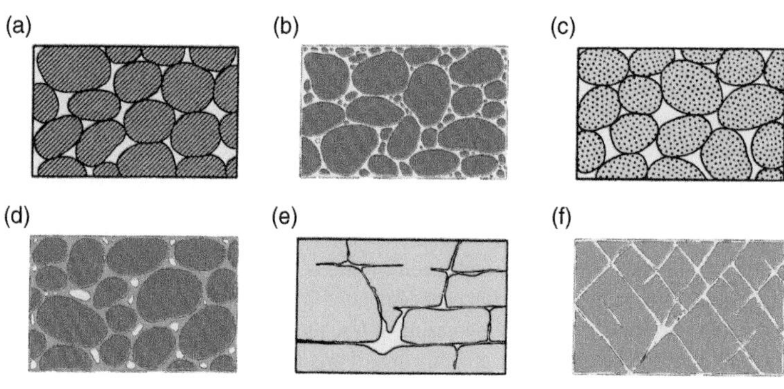

Figure 3.4 *Porosity in unconsolidated sediments varies with degree of sorting and shape of the grains. In consolidated rocks, porosity may be reduced by cement filling the pore spaces or increased by dissolution or by fracturing. See text for explanation of parts (a) to (f). (After Meinzer, 1923 / Public domain.)*

rock mass. Porosity is controlled by the grain size and shape, the degree of sorting, the extent of chemical cementation and the amount of fracturing. Figure 3.4 illustrates how porosity varies in unconsolidated sediments. The well-sorted sediments (a) have a high porosity that is reduced in (b), where the finer grains fill the pore spaces. In (c) the grains themselves are porous, thereby increasing the overall porosity of the deposit. In consolidated rocks the porosity is often reduced by the presence of cementing material (d). Fractured limestones may have the porosity increased by dissolution (e) or simply by fracturing (f). The amount of interconnected pore space that is available for fluid flow is termed the *effective porosity* and is also expressed as a percentage of the rock mass.

The porosity of solid rocks tends to be lower than their unconsolidated equivalents, as part of the pore space is taken up with cement. Some rocks with relatively high porosity values may be poor transmitters of water because the individual pores are poorly interconnected. Figure 3.4 illustrates some of the aspects of porosity development in consolidated rocks. Porosity that has developed after the rocks have formed is termed *secondary porosity* to distinguish it from *intergranular* or *primary porosity*. Secondary porosity typically results from two main causes: fractures associated with joints, along bedding planes, tectonic joints and faulting (although where fault gouge has been produced or secondary mineralisation has occurred along the fault plane, groundwater movement will be restricted rather than enhanced); and karst processes that dissolve limestone aquifers. Dolomitisation of limestones (i.e. the replacement of calcium ions with magnesium) also increases porosity because the magnesium ion is smaller than the calcium ion that it replaces by as much as 13%. However, the dolomite crystals are usually very small, producing tiny pore spaces, and are unevenly distributed through the rock, resulting in only small increases in the hydraulic conductivity.

Porosity does not provide a direct measure of the amount of water that will drain out of the aquifer because some of the water will remain in the rock, retained around individual grains by surface-tension forces. That part of the groundwater that will drain from the aquifer is termed the *specific yield*, and the part that is held in the aquifer is called the *specific retention*.

3.3 Hydraulic Properties and Rock Types

A great deal can be learned about the aquifer hydraulic properties from the study of aquifer geology, particularly its lithology and factors that influence fracturing. Most of the comments in this section relate to unconsolidated materials and sedimentary rocks that between them make up the vast majority of aquifers.

3.3.1 Porosity and specific yield

The proportion of porosity that makes up the specific yield is controlled by the grain size in non-indurated deposits. This relationship is shown in Figure 3.5 for unconsolidated aquifers. Specific retention decreases rapidly with increasing grain size, until it remains roughly around 6–8% for coarse sands and larger-sized sediments. Specific yield is at a maximum in medium-grained sands, because porosity decreases with increasing grain size. Note that a high degree of sorting will significantly reduce specific retention in coarse-grained sediments.

The ranges of values of porosity and specific yield for the common rock types are shown in Table 3.1. The relationship between porosity and specific yield is more complicated in solid rocks than in unconsolidated sediments, due to the effects of cementation and compaction in reducing the specific yield and the influence of fracturing in increasing it. Use this table to estimate an approximate value for specific yield for the aquifers in your study area. Porosity and specific yield values can also be estimated from the graph shown in Figure 3.5 based on the information you have obtained about the grain-size distribution for the aquifer. This graph is a best-fit curve based on scattered data and will only provide a rough estimate of the specific yield, so use the results with caution.

3.3.2 Permeability

Although porosity and permeability are affected by the same geological factors, these aquifer properties are different and should not be confused. Porosity is a measure of how much water the rock contains, whereas permeability determines how fast the water can flow through it. In this way, permeability and specific yield are broadly related so that, in general, aquifers that have a high specific yield tend to be more permeable, and less-permeable rocks usually have a lower specific yield.

Unconsolidated sediments generally tend to be much more permeable than their consolidated counterparts, because the cement reduces the overall void space in the rock, thereby reducing the

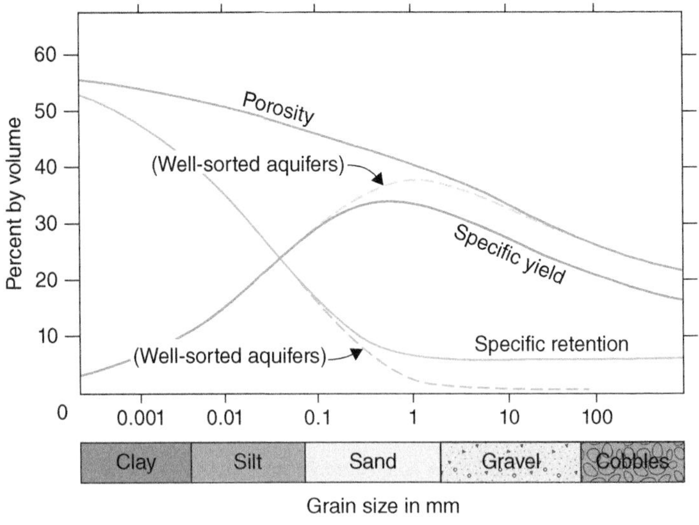

Figure 3.5 *Relationship between porosity, specific yield, specific retention and grain size applies to unconsolidated sediments only. Lines on these graphs are best-fit curves drawn through scattered points and are only approximate.*

Table 3.1 *Typical values of porosity and specific yield for a range of aquifer materials.*

Material	Porosity (%)	Specific yield (%)	Fraction of porosity producing specific yield
Coarse gravel	28	23	0.80
Medium gravel	32	24	0.75
Fine gravel	34	25	0.73
Coarse sand	39	27	0.69
Medium sand	39	28	0.7
Fine sand	43	23	0.53
Silt	46	8	0.17
Clay	42	3	0.07
Dune sand	45	38	–
Loess	49	18	0.37
Peat	92	44	0.48
Till (mainly silt)	34	6	–
Till (mainly sand)	31	16	–
Till (mainly gravel)	–	16	–
Fine-grained sandstone	33	21	0.54
Medium-grained sandstone	37	27	0.64
Limestone	30	14	0.47
Dolomite	26	–	–
Siltstone	35	12	–
Mudstone	43	–	–
Shale	6	–	–
Basalt	17	–	–
Tuff	41	21	0.51
Schist	38	26	–
Gabbro (weathered)	43	–	–
Granite (weathered)	45	–	0.80

Adapted from Water Supply Paper 1839-D (Column 1) and Water Supply Paper 1662-D (Column 2) by permission of the United States Geological Survey, and from Younger (1993) (Column 3) by permission of the Geological Society of London.

interconnection between pore spaces. As with porosity, the permeability of consolidated rocks will be increased by jointing and fissuring. This is termed *secondary permeability*. Rock types can be classified on the basis of having primary permeability, secondary permeability or both, as shown in Table 3.2.

Figure 3.6 provides hydraulic conductivity values for most unconsolidated sediments and rocks. This information can be used to estimate likely values for the aquifers in your study area. Again, the same words of caution must be repeated, as this will only be an estimate of the hydraulic conductivity and it must be used with care.

The relative permeability values of the different rocks within the study area will have a significant influence on the groundwater flow rates and directions, groundwater levels and the yield of wells.

Table 3.2 Classification of rock types in terms of permeability.

Type of permeability	Sedimentary		Igneous and metamorphic	Volcanic	
Intergranular	Unconsolidated Gravely sand, clayey sand, sandy clay	Consolidated	Weathered granite and weathered gneiss	Unconsolidated Weathered basalt	Consolidated Volcanic ejecta, blocks, fragments of ash
Intergranular and secondary		Breccia, conglomerate, sandstone, slate zoogenic limestone, oolitic limestone, calcareous grit		Volcanic tuff, volcanic breccia, pumice	
Secondary		Limestone, dolomite, dolomitic limestone	Granite, gneiss, gabbro, quartzite, diorite, schist, mica-schist	Basalt, andesite, rhyolite	

Major rock types that behave as aquifers have been classified on the basis of the type of permeability that they exhibit. Intergranular or primary permeability is a feature of unconsolidated deposits and weathered rocks. It also occurs in most sedimentary rocks and those igneous rocks that have a high porosity. Secondary permeability is largely due to fissuring or solution weathering and only affects indurated rocks.

Adapted from *Groundwater in the Western Hemisphere* (1976) by permission of the United Nations.

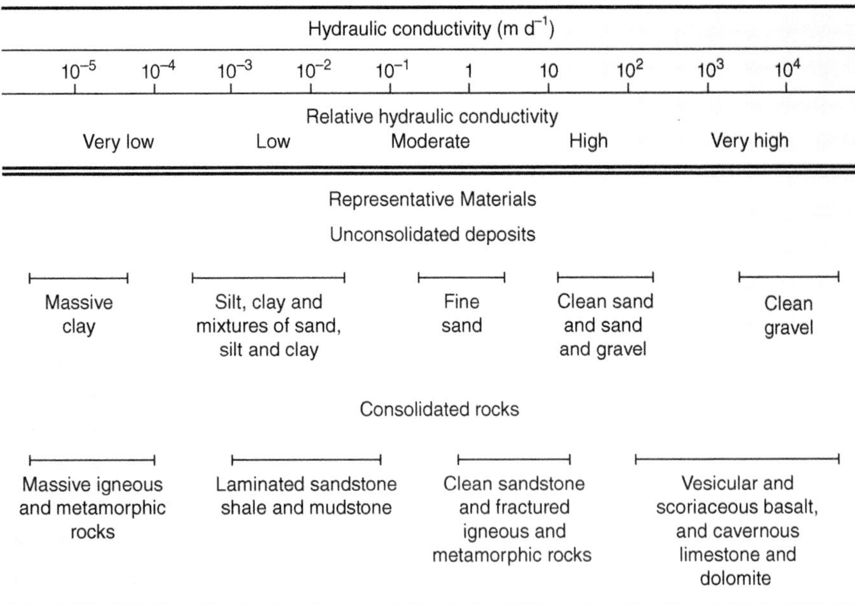

***Figure 3.6** Hydraulic conductivities in metres per day for various rock types. (Ground Water Manual/United States Department of the Interior/Public domain.)*

Form a mental picture of the way that both groundwater flow rates and the slope of the water table vary in each rock type to understand the groundwater system you are studying. Instead of thinking in terms of hydraulic conductivities you may find it easier to use the concept of *hydraulic resistance*, the opposite of conductivity, and this can be thought of as $1/K$.

Hydraulic resistance is not a common concept in hydrogeology, although it is widely used in aerodynamics and to evaluate fluid flow through pipes. I find it helpful in visualising groundwater flow through large rock masses that have a variety of different permeabilities, as it seems easier to associate reduced flow with increased resistance and greater flow with lower resistance than with a comparison of varying conductivities. Good aquifers have a low resistance to flow and develop relatively shallow hydraulic gradients. Poor aquifers and aquitards, on the other hand, present a significant resistance to groundwater flow and are associated with steep hydraulic gradients for similar flow rates. Use the tables of hydraulic conductivity values to work out the relative hydraulic resistances of each rock type.

It is possible to estimate the hydraulic conductivity for sandy sediments from their grain size using the Hazen method (Hazen, 1911). Hazen developed his formula from empirical studies carried out to improve the design of sand filters for drinking-water treatment. The method uses grain-size distribution information obtained from sieve analysis and can be used to obtain a general idea of the hydraulic conductivity of sands. Warning – do not try to apply it to other materials! Plot the particle distribution curve for a number of aquifer samples to obtain an average value for the grain size at the 10% value (see Figure 3.1b), which was termed the *effective grain size* (D_{10}) by Hazen. The *Hazen formula* is:

$$K = C(D_{10})^2$$

In this formula, K is the hydraulic conductivity in metres per day, D_{10} is the effective grain size in millimetres and C is a dimensionless coefficient taken from Table 3.3. Use your geologist's judgement to select a value for C based on the material type. For example, a clayey fine-grained sand would have a value of about 350 for C whereas a medium- to coarse-grained sand would have one of around 850.

Hydraulic conductivity has an incredibly wide range of possible values. Those shown in Figure 3.6 are on a logarithmic scale; hence, the hydraulic conductivity of clean gravel can be expected to be around one thousand million (10^9or a billion in the USA) times greater than that of massive clay (i.e. nine orders of magnitude). In fact, it is quite common for hydraulic conductivities to be quoted to the nearest order of magnitude (e.g. 10^{-1} m d^{-1}), rather than as a precise value. This approach reflects the reality of spatial variations within an aquifer caused by geological factors, which basically means that it is virtually impossible to measure it with complete accuracy for an aquifer as a whole.

These variations in an aquifer's permeability are taken into account in the concept of transmissivity being the product of the hydraulic conductivity and the full thickness of saturated aquifer. In practice, the 'effective' thickness is used rather than the total thickness, especially in thick and/or

Table 3.3 *Values of C in Hazen's formula.*

Grain size	Values of C
Very fine sand – poorly sorted	350–700
Fine sand with fines	350–700
Medium sand – well sorted	700–1000
Coarse sand – poorly sorted	700–1000
Coarse sand – clean and well sorted	1000–1300

Values given only apply when K is in metres per day and D_{10} is in millimetres.

anisotropic aquifers. In many instances, the effective thickness is taken as the depth of the boreholes used to measure the transmissivity value. From this definition it follows that the metric unit of transmissivity is square metres per day, a simplification of $m \times m^3\ d^{-1}\ m^{-2}$. The permeability values obtained from pumping tests (see Chapter 6) are usually expressed in terms of the aquifer's transmissivity. More than 50 years ago, the term *transmissibility* was used rather than transmissivity and has the same meaning.

3.3.3 Dispersion

The way in which a body of water is dispersed as it flows through an aquifer is related both to the type of porosity and to the aquifer characteristics. Groundwater molecules move at different rates depending on whereabouts in the aquifer they are flowing. This is caused by friction on pore walls, variations in pore sizes and variations in the path length. As groundwater flows through the pores, it will move faster at the centre of the pore than along the walls due to friction. Pores usually vary in size, and so groundwater will flow through larger pores faster. Some flow paths are more tortuous than others; consequently, some molecules will travel along longer pathways than others. Because groundwater is not all moving at the same rate, mixing occurs along the flow path. This mixing is termed *mechanical dispersion*. The mixing that occurs along the direction of groundwater flow is termed *longitudinal dispersion* and the mixing that occurs normal to the direction of fluid flow is termed *transverse dispersion*.

It is easier to think about dispersion by considering a tracer injected into an aquifer. The tracer will not retain its original volume because mechanical dispersion will cause it to be diluted (Figure 3.7). This property of aquifers is important in tracer work (see Section 9.3) and is fundamental to the consideration of the movement of pollutants through an aquifer.

3.3.4 Storage

The volume of water in an aquifer is calculated as the *storativity* or *storage coefficient*, which has the same value as the specific yield in water table aquifers but not under confined conditions. The storativity is defined as the volume of water released from a unit volume of aquifer for a unit decline in head. When water is pumped from a confined aquifer, it causes a lowering of water levels, but this represents a reduction in pressure and not the aquifer being drained. It is analogous to letting air out of a car tyre, where a measurable volume of air is removed but the tyre remains full, albeit at a reduced pressure. Storativity of confined aquifers lies in the range 0.00005–0.005, indicating that

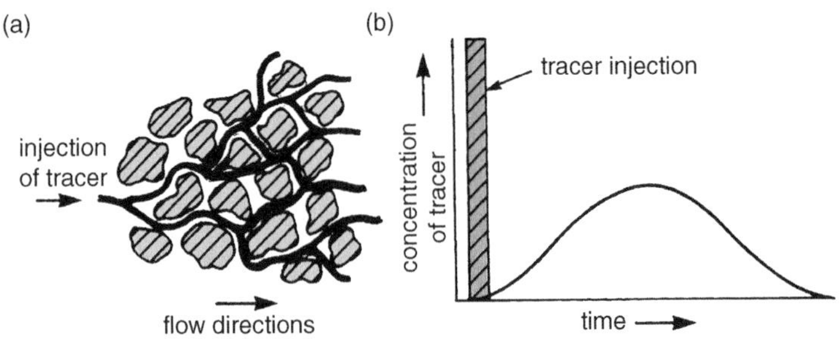

Figure 3.7 *As a tracer flows through a rock (a), it divides each time an alternative pathway is reached. This process dilutes tracer concentration by mechanical dispersion and mixing. Graph (b) shows how concentration of a tracer varies with time as it flows past a particular point in an aquifer or emerges at a discharge point.*

substantial changes in pressure are required over large areas to produce significant yields of water. This difference between the storage coefficients of confined and water table aquifers is illustrated in Figure 3.8. In a water table aquifer (a), unit decline in head produces a volume of water equivalent to the specific yield by dewatering a unit volume of rock. With a confined aquifer (b), the unit decline in head releases a relatively small drop of water and the aquifer remains fully saturated.

3.4 Assessing Hydraulic Properties

You should now be able to estimate likely values for the hydraulic conductivity and specific yield for each rock type in the study area and classify them as potentially good aquifers, poor aquifers, and non-aquifers. You will also be able to decide whether the permeability is essentially primary, secondary, or a mixture of both. Even in areas where good geological maps are available, you may not be able to do this without going into the field, unless you already have a good knowledge of local lithologies and structure.

During your walkover survey, look at the grain sizes and degree of sorting of sediments and sedimentary rocks. Use a grain-size chart to determine the grain sizes and carefully record the location of each reading. If necessary, bring samples back with you and have them tested or test them yourself. You can use the Hazen formula to estimate the permeability of sandy materials. Inspect exposures of solid rocks for joints and other fractures and, if possible, examine cores recovered from boreholes in the area, to see if the rocks change with depth. Remember that in almost all instances the size of individual fractures seen in rock faces is greater than that which occurs in the main body of the rock. Stress release takes place when the rock face is exposed by quarrying or erosion, and this allows the fractures to widen. In aquifers where fracture flow is likely to be important, record the trend of joints and identify the direction of joint sets. This will help determine the possible

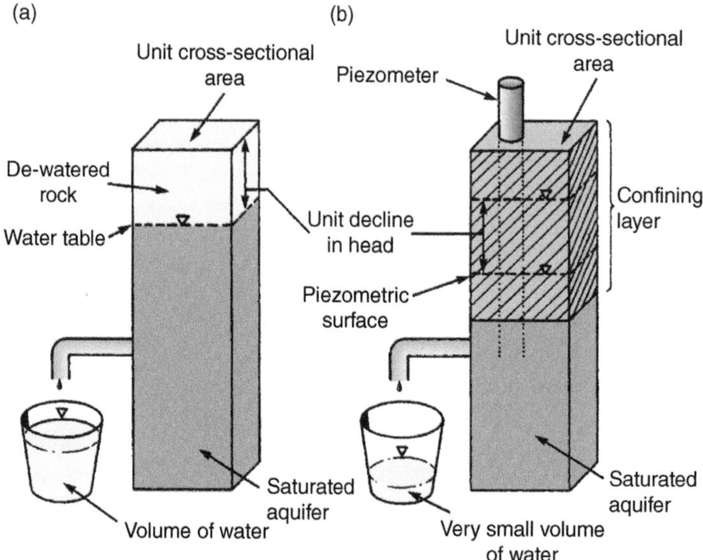

Figure 3.8 *The concept of storativity. (a) Unit decline in head in a water table aquifer releases a volume of water equivalent to the specific yield. Under confined conditions (b), the unit decline in head releases a very small volume and the aquifer remains fully saturated.*

directions of groundwater flow. Your geological training should be adequate to enable you to complete these field exercises without undue difficulty. As information is gathered about the aquifers, draw it together in a summary form, as in Figure 3.9. Use the information contained in this chapter to estimate likely values of aquifer properties from which you can identify the main aquifers. Supplement this information with notes on relevant points. These notes are used in conjunction with a geological map of the area and groundwater-level information to complete the picture of the groundwater-flow system. You will now be able to use the estimated values of the aquifer properties and other information to help interpret groundwater-level data and construct groundwater-level contour maps and flow nets.

More precise values of hydraulic conductivity and specific yield can be measured from a range of field tests that involve inducing a flow of water through the aquifer and measuring the change in water levels that result. These tests form a subject in themselves and are described in Chapter 6.

3.5 Using Hydraulic Property Information

Estimates of the hydraulic properties will help identify those aquifers likely to give the highest yields and will help you to design a new well. Conversely, you may be looking for a suitable location for a landfill site and need an area with a very low hydraulic conductivity and no secondary permeability. The information contained in a summary such as shown in Figure 3.9 will enable you to decide which areas are worth more detailed consideration. In this example, any of the three main

Aquifer type	Grain size, sorting, etc.	Estimated hydraulic conductivity (m d^{-1})	Estimated specific yield (m d^{-1}) (%)	Type of permeability	Notes
Main aquifers					
1. Glacial sand and gravel	Medium/coarse sands and fine gravel with some cobbles	10–10^2	25	Primary	Grain-size analysis
2. Triassic sandstone	Fine/medium sandstone, well cemented in parts	1–10	15	Primary + secondary fractures	Confined by till in some parts
3. Carboniferous limestone	Massive dense limestone with joint related fractures	10^2	15	Secondary via joint fractures	Some evidence of karst features
Poor aquifers					
1. Alluvium	Mainly silt and thin sands	10^{-3}	5	Intergranular	Limited to valley bottoms
2. Granite	Joints opened by weathering to 2 m depth	10	5	Secondary via fissure	Joint sets mappped
Non-aquifers					
1. Glacial clay	Mainly clay, some silt	10^{-5}	< 5	Primary	Till and varved clays
2. Carboniferous mudstone	Mainly mudstone, some siltstone, well bedded	10^{-4}	< 1	Secondary in weathered rock	Slight seepage seen on some bedding planes

Figure 3.9 *Build up a picture of the aquifers in your study area based on the information you have collected or estimated using methods described in the text.*

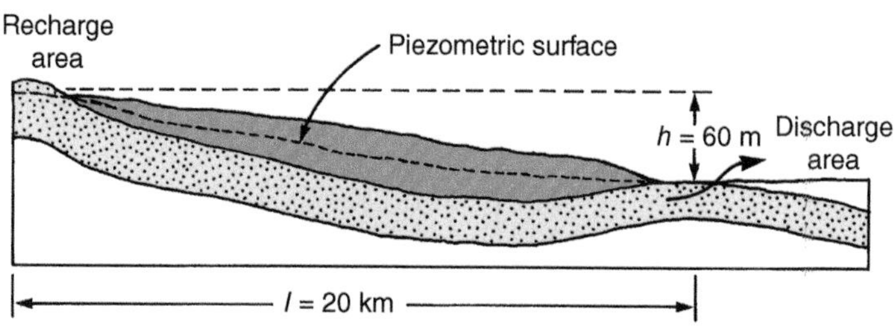

Figure 3.10 *Regional flow (Q) through a sandstone aquifer can be calculated using Darcy's law.*

aquifers would ensure high-yielding wells, but the success of a well in the Carboniferous Limestone would depend on the extent of the fracture system encountered in the well, as all the groundwater flow is through these fractures. Possible locations for landfills may be found on either the glacial clay or the Carboniferous mudstone, although there is a risk that secondary permeability along bedding planes in the mudstones may allow polluted water to escape.

You can assess the overall flow through an aquifer by applying Darcy's law, as shown in Figure 3.10. The sandstone has an average thickness of 200 m and is 10 km wide. The distance from the recharge area to the discharge points is 20 km, and the head difference is 60 m on average. The hydraulic conductivity is 5 m d^{-1}. Substituting these values into Darcy's law (see Section 3.2) we get

$$Q = kAh/l = 5 \times (200 \times 10\,000) \times 60/20\,000 = 30\,000 \text{ m}^3\text{d}^{-1}$$

Assuming the effective porosity (α) is 0.2, the velocity (v) can be calculated as follows:

$$v = \frac{Kh}{\alpha l} = \frac{5 \times 60}{0.2 \times 20\,000} = 0.075 \text{ m d}^{-1}$$

This means that it would take some 712 years for water to travel the 20 km from the recharge area to emerge from springs in the discharge zone under the prevailing head conditions. The calculation assumes both uniform groundwater flow and a homogenous aquifer, which in the context of a regional system are valid.

3.5.1 Pumping from wells

When a well is pumped, the water level around the well falls in response to the pumping, forming a *cone of depression* (Figure 3.11). The shape and extent of the cone of depression depends on the rate of pumping, the length of time pumping has continued, and the hydraulic characteristics of the aquifer. The amount that the water table has been lowered is called the *drawdown*. One of the many formulae that relate these parameters is the *Thiem* or *equilibrium* or *steady-state* well equation:

$$Q = \frac{K(H^2 - h^2)}{C \log(R/r)}$$

where Q is the pumping rate, K is the hydraulic conductivity, H is the thickness of the saturated aquifer penetrated by the well, h is the height of water in the well, r is the radius of the well, and R is the radius of the cone depression. C is a constant with the value of 0.733; Q is in cubic metres per day, K is in metres per day, and the other dimensions are in metres.

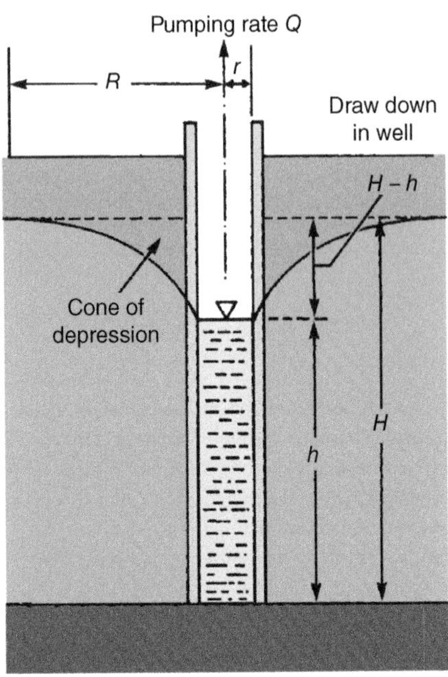

Figure 3.11 *The water table is drawn down into a cone of depression around a pumped well. Q, R, r, H and h are used in the equilibrium well equation (see text) to calculate hydraulic conductivity of the aquifer.*

This equation can be used to estimate the yield of a well with different amounts of drawdown once they have been stabilised, if the hydraulic conductivity is known, or it can be used to calculate the hydraulic conductivity if the drawdown and pumping rates are measured once the groundwater levels have ceased to change.

A graphical method that uses steady-state data to calculate transmissivity (hydraulic conductivity × the aquifer thickness) was proposed by Thiem (1906) with the following equation:

$$T = \frac{2.30Q}{2\pi\,\Delta s}$$

where:

T is transmissivity in square metres per day;
Q is the discharge rate in cubic metres per day;
Δs is the slope of the straight line distance–drawdown graph as the drawdown over one log cycle on the time axis.

The drawdown is plotted against distance on a log-scale and a best-fit straight line drawn though the data points. The slope of the graph is substituted into the equation shown in Figure 3.12 to calculate the transmissivity. During a pumping test on a new borehole, water level measurements were made at an observation borehole 250 m from the test well. At the end of the test the drawdown in this borehole was 2.5 m. No close-proximity borehole was available, and so Bierschenk's method (see Figure 6.11) was used to estimate the drawdown in the aquifer adjacent to the pumping

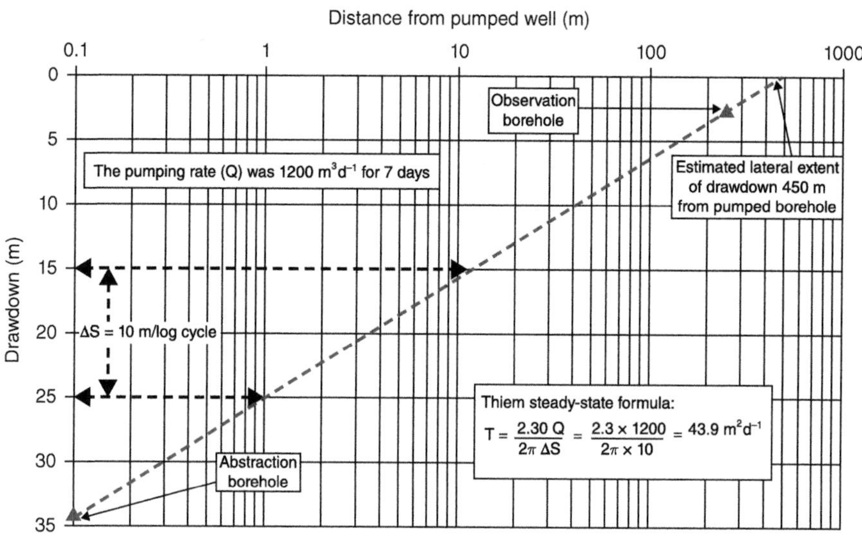

Figure 3.12 *Thiem's steady-state equation can be solved graphically, as shown in this example.*

borehole. Thiem's graphical method was used to calculate the transmissivity as approximately 44 $m^2 d^{-1}$. The boreholes penetrate 65 m of aquifer, making the hydraulic conductivity about 0.68 m d^{-1}, which is at the low end of the typical range for sandstone aquifers (see Figure 3.6). The slope of the line also indicates the radius of influence of the pumped borehole as about 450 m.

Logan (1964) suggested a simplified steady-state equation relating the drawdown in a pumping well at a given rate of pumping to the aquifer's transmissivity.

Logan's equation is:

$$T = \frac{1.22Q}{s_w}$$

In this equation, T is the transmissivity in square metres per day, Q is the pumping rate in cubic metres per day, and s_w is the drawdown in the pumping well in metres. In fact, the equation will work in any units provided that you are consistent and do not mix them. For example, do not use a pumping rate in seconds or hours when the time dimension in transmissivity is in days!

Logan's method will only give a very approximate value for transmissivity (and hence hydraulic conductivity) because it uses the drawdown in a pumping well, which does not truly reflect the drawdown in the aquifer, as pumping drawdown is increased by well losses. However, with care it can be used to extend your understanding of an aquifer where more sophisticated data are not available. More information on obtaining and using better quality data from pumping tests is given in Chapter 6.

3.5.2 Sensitivity analysis

When making an assessment of groundwater flow or storage based on an estimated value for the aquifer's hydraulic properties you should test the significance of the possible variations in these values. This technique is called *sensitivity analysis* and consists of repeating the calculation using the maximum and minimum values within the possible range and comparing the different results.

This approach will show how critical values of a particular parameter are to making predictions with an acceptable reliability. For example, if you are assessing the potential for a new abstraction to lower the water level in existing wells, you may use the steady-state equation to calculate the draw-down at certain distances from the new well. If you use the possible maximum and minimum values for the aquifer hydraulic conductivity and the lower value shows that problems may occur, then it is essential for you to derive a more precise value of the hydraulic conductivity to make a more reliable prediction.

3.6 Recharge through the Vadose Zone

Groundwater is mainly recharged by rain falling on the ground, then soaking into it and flowing through the soil zone and then on through the vadose zone (or unsaturated zone) until it reaches the water table. This part of the hydrological cycle usually gets only limited attention from hydrogeologists because their main interest is in saturated aquifers.

Just over 40 years ago I led a team of hydrogeologists looking at the increasing nitrate content in groundwater to identify its rate of increase and one of the things we needed to determine was the speed with which water was passing through the vadose zone.

At that time the atmosphere contained high concentrations of tritium that were a by-product of thermonuclear bomb testing in the atmosphere that started when the US exploded its first atomic bomb on 16 July 1945 at a desert test site in Alamogordo, New Mexico. These tests were banned by the 1963 Limited Test Ban Treaty that most countries followed meaning that the atmospheric concentrations went down from the mid-1960s.

Tritium (^3H) is a radioactive isotope of hydrogen with a half-life of about 12.32 years. The nucleus of tritium contains one proton and two neutrons, whereas the nucleus of the most common isotope of hydrogen, protium (^1H), that forms 99.98% of all hydrogen isotopes, contains one proton and zero neutrons. That of deuterium (^2H) contains one proton and one neutron. Neither protium nor deuterium are radioactive. Naturally occurring tritium is extremely rare on Earth with the atmosphere normally only having trace concentrations (<5 TU) (tritium units) formed by comic bombardment of atmospheric gases, it is a by-product of nuclear bomb testing that gave atmospheric concentrations of more than 6,000 TU in 1964, its maximum concentration. The atmospheric concentration declined slowly as tritium decayed and it was possible to use it as a groundwater tracer until around the early 1990s.

As tritium is a form of hydrogen, it readily takes the place of protium or deuterium in water molecules. It falls to the Earth in rain or snow and will infiltrate into the ground within the recharging water to flow through the vadose zone. It is therefore, the best groundwater tracer that there can be because tritiated water moves exactly where non-tritiated water flows. For this reason, back in the day, we used atmospheric tritium to detect the rate of flow through the vadose zone in the Sherwood Sandstone aquifer in northwest England; similar work was also carried out by other water authorities and by research bodies across the country such as the British Geological Survey and embraced all the major British aquifers (Smith et al., 1970).

The process was to drill a borehole using air as the circulating fluid, taking continuous 200 mm diameter core. In the majority of cases these boreholes were intended to become observation boreholes and were drilled to final depths of 120 m to 180 m with the water table being at 10–50 m below ground.

Pieces of core about 30 cm long were taken every couple of metres or so down the borehole in the section above the water table. The core was then wrapped in aluminium foil, the foil was sealed with tape, and a label stuck on the foil identifying the core depth and the borehole location. It was then placed in a plastic bag that was also sealed and labelled with the same information. Finally, it was put into a second plastic bag that was also sealed and labelled and it was then placed into a freezer. This packaging and excess labelling were found to be necessary as some of the labels and even some of the wrapping could be lost when freezing large pieces of rock.

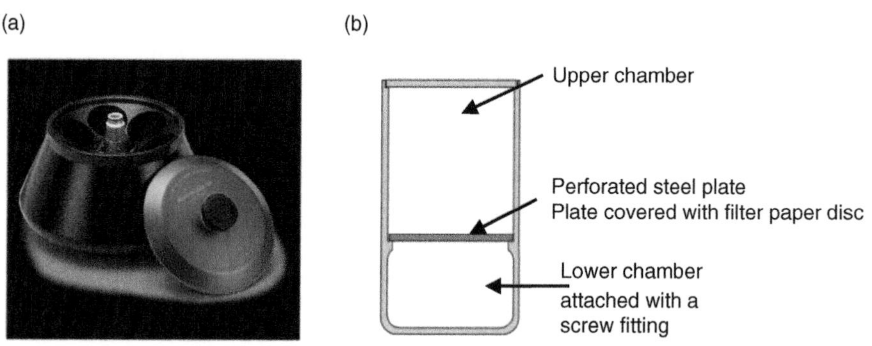

Figure 3.13 *Centrifuge head and cup (a) Centrifuge head (courtesy of Beckman Coulter Ltd.) (b) typical centrifuge cup made of a plastic with a steel inner plate.*

The freezing kept the water contained in the core in place until it was recovered. This was done by allowing the core to thaw for a period of a few hours and then the outer 25–30 mm was removed using a chisel and a 4 lb hammer. It was then reduced to sand with some silt and clay by being pounded first by the hammer, and then using a pestle and mortar. The sand was placed into six plastic high-speed centrifuge cups making sure that each one weighed the same. Each cup had an upper compartment that had at its base a stainless steel plate with a large number of holes penetrating it. The sample sat in the upper chamber on a disc of filter paper.

It was then spun at a high speed for a set period until the water was separated from the sand by centrifugal force and was collected in the lower chamber of the cups (see Figure 3.13). The water was then poured into a sample bottle. These bottles were collected and then sent to a specialist laboratory for the tritium content to be measured. When the results were returned a vertical plot was made that showed the variations in tritium concentration with depth.

The exercise was repeated a year or two later with the second borehole being drilled closeby to the first one. The results from the second exercise were also plotted on the depth profile where the maximum concentration was easily seen and illustrated by Figure 3.14. The difference in depth between the two peaks was then measured and the vertical velocity for percolation through the rock was calculated using the difference in depths and the time between the measurements. The results were summarised in a paper by Chilton and Forster (1991) with these data reproduced as Table 3.4.

The flow velocities through the Chalk are shown to be 0.3 – 1.4 m/year, supporting calculations previously made by Grindley (1969) who used conventional meteorological models to calculate flow velocities in the vadose zone. In areas where the Chalk is highly fissured and areas of fissured Jurassic Limestone, the vertical flow is dominated by flow through fissures resulting in the tritium peak not being well preserved. In these areas the *bypass flow* or fast flow dominates and the unsaturated residence times are at the low end of the reported ranges. The bypass flow rates are controlled by the fracture sizes, the length of the flow path and its tortuosity allowing the time taken to vary from a few hours to about ten days or so, to reach the water table.

The element of flow through unsaturated rock where it is contained in the pore spaces is described as *piston flow* which is where each recharge event enters at the top of the partially saturated vadose zone and forces a similar volume of water out at the bottom. This is the dominant flow method through unsaturated deposits in granular aquifers after a period that ranges up to around 10 days. The degree to which bypass flow occurs is related to the presence of fractures, which is related to the extent to which the rock body is cemented. Moderately fissured rocks will have both piston flow and bypass flow operating at the same time and hence, part of the recharge water flows down to the water

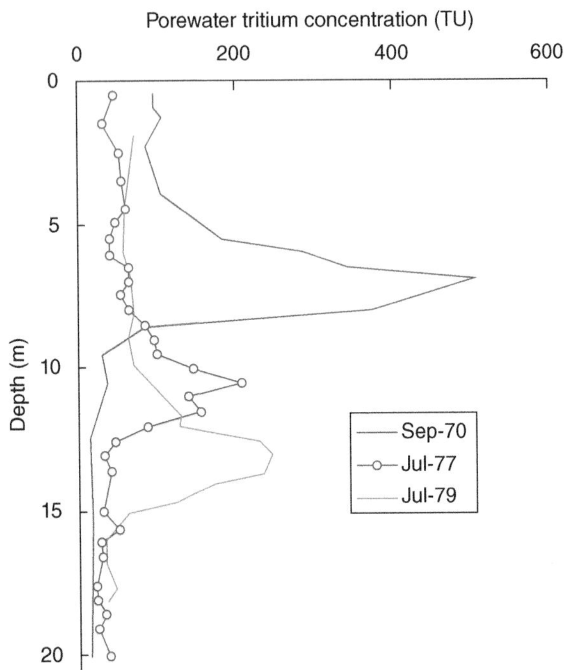

Figure 3.14 *Tritium profiles (from Stuart et al., 2016/British Geological Survey).*

table in a number of years through piston flow and some of it races down in a matter of days as bypass flow. The proportion of the total flow in each system will depend on the extent of the fracture system that allows the bypass flow to take place. Therefore, the recharge processes provide a natural system of mixing the different recharge waters to a considerable extent. These results provide a model of flow through the vadose zone where piston flow is the natural method of recharge through uncemented and poorly cemented rocks, and where fractures exist recharge also takes place through them.

The difficulty with the piston flow theory is an approximately equal volume of water is pushed out of the bottom of the vadose zone with a relatively small time delay. This allows an apparent correlation between the recharge and observed changes in groundwater levels or stream base-flows that can produce a good match although assuming that the events are actually linked is clearly mistaken. Although these recharge processes were defined more than 30 years ago there are still hydrogeologists who do not believe them despite the irrefutable evidence of the movement of the tritium peak.

The United States Geological Survey reported in 2010 on the results of sites in the USA where piston flow values had been calculated during the period 1992 to 2005 (Hinkle et al., 2010). These measurements used the different tracers referred to in this section although they are not described in detail as these are very specialist methods. Those who wish to pursue the methods use should start with the online publication by Hinkle *et al.*, (2010) and use the reference lists as a guide.

While we are thinking about the flow rates through the vadose zone it is worth considering why they are so very much slower when compared with the flow of groundwater through the saturated zone of an aquifer. The vadose zone is not fully saturated with water and indeed the amounts of water it contains vary with time. This means that the flow takes place through only part of the rock which varies with the time since the last recharge event.

Table 3.4 Showing results of recharge investigations using tritium (from Chilton and Foster, 1991/Springer Nature).

	Cretaceous chalk	Triassic sandstone	Jurassic limestone
Lithology	Microporous carbonate	Quartz grains with sparse silica or carbonate cement	Microporous or dense carbonate
Ground-water flow regime	Fissure and matrix	Intergranular with some fissure	Fissure
Matrix porosity (%)	25–45	15–35	10–25
Characteristic pore size (μm)	0.2–1	5–50	0.05–0.5
Matrix hydraulic conductivity (m/d)	2×10^{-4} to 5×10^{-3}	5×10^{-1} to 10	up to 5×10^{-4}
Effective rainfall (mm)	150–350	200–350	150–250
Unsaturated zone flow rates (m/yr)	0.3–1.4	0.6–2.3	0.6–2.5
Unsaturated zone thickness (m)	5–50	10–40	5–25
Unsaturated zone residence time (yr)	4–120	3–65	2–40
Fissure spacing (m)	5–10 (macro) 0.1 (micro)	Very variable and difficult to generalise	1–5 (macro)
Natural ground-water flow velocity (m/d)	1–10		5–25
Aquifer thickness (m)	Up to 450	Up to 500	20–35

The hydraulic conductivity of the rock that controls the flow rate is a product of the rock properties and the amount of water present. The vast majority of times when the hydraulic conductivity is assessed, data from pumping tests are used to calculate it to give a saturated value and which does not apply to the partly saturated vadose zone. The value of hydraulic conductivity in the unsaturated zone varies greatly from just below the value obtained from pumping tests to zero when there is no water present other than that held round rock particles by surface tension forces.

The graph below (based on Selker and Or, 2022) represents the changes in the hydraulic conductivity value in relation to the soil moisture content in the vadose zone. The hydraulic conductivity at the top right of the graph is that of the saturated aquifer and is determined from pumping or other tests.

The relationship between moisture content and hydraulic conductivity forms a hyperbolic line rather than being a straight line to zero. It does not reach the point where soil moisture is zero because of the residual water content. Cohesion holds the hydrogen bonds together to create surface tension in water. Since water is attracted to other particles such as those that make up the rock, adhesive forces pull the water towards them. The residual water (Θ_r) is retained on the surface of rock particles when the surface tension is greater than the force of gravity, which occurs when the total volume of water is approaching zero.

The diagram to the left in Figure 3.16 (1) represents grains that form the rock with pore spaces between them. At point 'A' in Figure 3.15, the rock is fully saturated allowing water to flow through all the pore spaces that are available. The second diagram (2) represents conditions at point 'B' in Figure 3.15 where the rock has been drained with some water remaining in a few locations in the network of pore spaces. This water is represented by light blue shading and the other pore spaces are

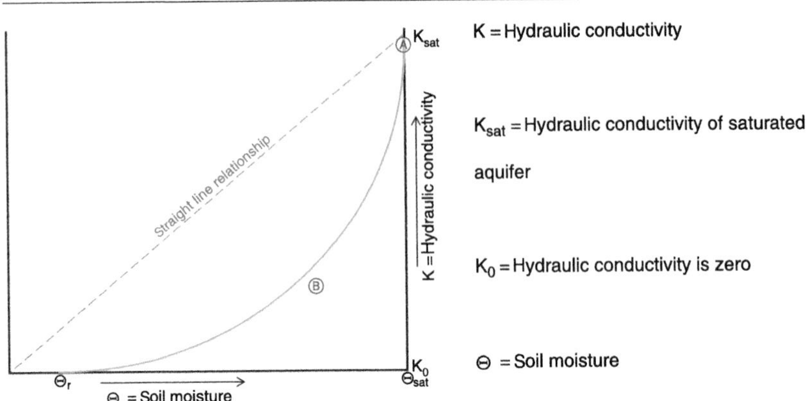

Figure 3.15 *The relationship between hydraulic conductivity and water content of the rock (based on Selker and Orr, 2022).*

dry. As a result, the flow paths are longer and have an increasingly smaller cross sectional area. The pathways become ever more narrow as the volume of water reduces and the water flows along as increasingly thinner films of water. These factors result in the hydraulic conductivity being much lower – that is the resistance to flow is much greater – than the values that would fall on the straight line from K_{sat} to zero as shown by the blue line in Figure 3.15. This explanation accounts for the lower velocities in partially saturated rock as the water content is reduced.

When water drains from the rock it follows that the larger pore spaces will dry out first. As a result, the water flow becomes concentrated in ever decreasing sized pores. The resistance provided by the pores to water flowing through it is proportional to the radius of the pore space to the fourth power (r^4). This means that for example, the flow through a 0.1 mm radius pore is 10,000 times smaller than the flow through a pore space with a 1 mm radius (ten-times bigger) simply by considering the resistance to flow. However, as the pores become smaller, more pores can fit into a given volume of rock. The rise in the number of pores increases the volume flowing through them by 100-fold (r^2) so the difference between a 1 mm radius pore and a 0.1 mm radius pore in a rock is actually only 100 times smaller. As a result of all these factors discussed here, the hydraulic

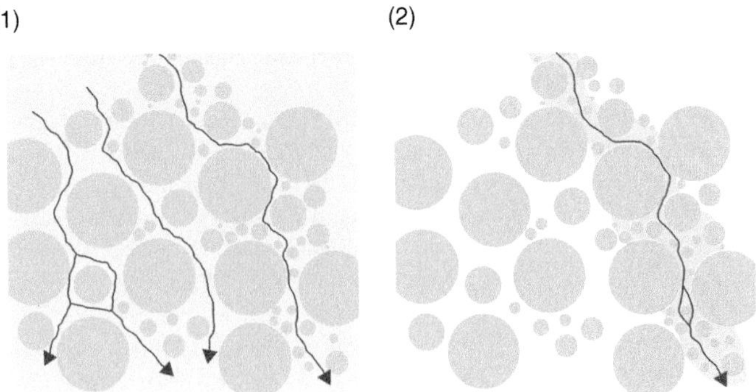

Figure 3.16 *Changes in flow path resulting from variations in the degree of saturation. (1) Flow through fully saturated rock. (2) Flow through rock unsaturated rock.*

conductivity in the unsaturated zone becomes smaller at the rate indicated by the curve in Figure 3.15 rather than the much greater rate that is suggested by the straight line on the graph.

So in summary, the values of the hydraulic conductivity of an aquifer that are usually quoted will have been calculated from pumping tests carried out using boreholes that penetrate the saturated rock. In the vadose zone the hydraulic conductivity varies hugely, from close to the value calculated from pumping tests down to zero when the only water in the aquifer rock is held to the rock particles by cohesive forces and cannot move. Well, the vadose zone is not also called the unsaturated zone for nothing!

Since the atmospheric tritium content fell to low levels making it difficult to use as a tracer other methods have been developed to assess flows in the vadose zone and for other forms of groundwater studies. Those of interest are chlorofluorocarbons (CFCs), sulphur hexafluoride (SF_6), and tritium/helium-3 ($^3H/^3He$). Chlorofluorocarbons (CFCs) are stable, synthetic, halogenated alkanes that are in a gaseous form and were developed in the early 1930s for use in refrigeration with other CFC compounds being developed subsequently. These compounds leak into the atmosphere and can be dissolved in the recharging waters and so may be used as tracers and dating tools for younger waters (post 1940). By measuring concentrations of CFC-12, CFC-11 and CFC-113, it is possible to identify groundwater recharged since approximately 1941, 1947, and 1955, when these compounds were released into the atmosphere respectively. Groundwater dating with CFC-11, CFC-12, and CFC-113 is possible because the atmospheric mixing ratios of these compounds are known and/or have been reconstructed from 1940 to the present; the Henry's law solubilities (proportional to its partial pressure) in water are known; and the concentrations in air and young water are relatively high and can be measured (Henry, 1803).

SF_6 is a trace atmospheric gas that is primarily produced artificially although it also occurs naturally in fluid inclusions in some minerals and igneous rocks, and also in some volcanic and igneous fluids. SF_6 is used as a dating tool of post-1990s groundwater because, unlike the chlorofluorocarbons with steady or declining atmospheric mixing ratios, atmospheric concentrations of SF_6 are expected to continue increasing (Busenberg and Plummer, 1997). SF_6 is also stable in reducing groundwater environments, and because there are relatively few uses of the compound, few environments are contaminated by artificial sources.

The tritium/helium-3 technique has become increasingly important over the last decade or so as methods using just tritium had become less useful. Helium is a noble gas, which means it is stable and does not have chemical reactions with other elements found in rocks or water making it a consistent and reliable reference point. By knowing the concentration of the helium isotope that comes from tritium as it decays radioactively (helium-3) compared to the total helium in the water, as well as the concentration of other noble gases, it is possible to determine the exact age of young water.

4

GROUNDWATER LEVELS

A good set of reliable groundwater-level measurements is the best foundation on which to build an understanding of a groundwater system. This chapter describes the instruments and equipment used to measure groundwater levels and how to use them, as well as the network of wells, boreholes and piezometers needed to make these measurements and how to manage and interpret the data.

Tables B1 to B7 in Appendix B provide conversion factors for changing units from any type used in collecting groundwater data to the type of units that you may use in your studies. These are particularly useful when converting long data sets that may have been started several decades ago and used units that are different to the metric units that are in common usage today.

4.1 Water Level Dippers

Among the hydrogeological fraternity, any instrument that is lowered into a borehole or well to measure the water level is called a 'dipper', and although this may be a jargon term, I have used it throughout this book. Most dippers consist of a flat tape graduated in metres and wound on to a drum. Wires run down each side of the tape and are attached to a probe that senses the water surface by completing a circuit and causing a buzzer to sound. Figure 4.1 shows the main features of such a typical dipper. Most commercially available dippers are made on the same general format. Batteries (usually totalling 9 volts) are housed in the drum spindle, which also contains electronic circuitry. The probe is made of stainless steel and acts as an electrode, with a second inner electrode being a pointed steel rod visible through holes in the weighted end. The electrodes are connected by a pair of wires that are incorporated into each edge of the plastic tape. When the point touches the water surface a circuit is completed and a buzzer sounds.

This equipment allows quick and easy readings to be taken, and the tape can be read to the nearest half centimetre if required. Some dippers have tapes graduated in millimetres, although field measurements to this accuracy are rarely achieved in practice. Dippers come in a wide range of sizes, from a compact 10 m up to an enormous 500 m length! When purchasing a dipper, ensure that it is long enough to measure the deepest water table in your area, bearing in mind that pumping-water levels in abstraction boreholes are much deeper than rest-water levels. A dipper 100 m long will cater from most situations, but 200-m tape may be required for measuring some pumping levels in boreholes with large drawdowns. If you are studying a shallow aquifer where there are only hand-dug wells, a 10-m dipper should be adequate. It is important to make these considerations as the size of the dipper affects the cost, and if you can get away with a small one they are very much more convenient to carry and use. Some modern dippers also incorporate other measurements such as electrical conductivity and temperature.

Some dippers have round section cable, with the depth graduations marked in metres by adhesive bands. It is necessary to use a steel tape to measure the distance in centimetres (i.e. to the nearest 0.01 m) from the nearest metre mark to obtain a more precise water level reading. These instruments are not easy to use and are hopeless for pumping tests, where frequent measurements are required.

Field Hydrogeology, Fifth Edition. Rick Brassington.

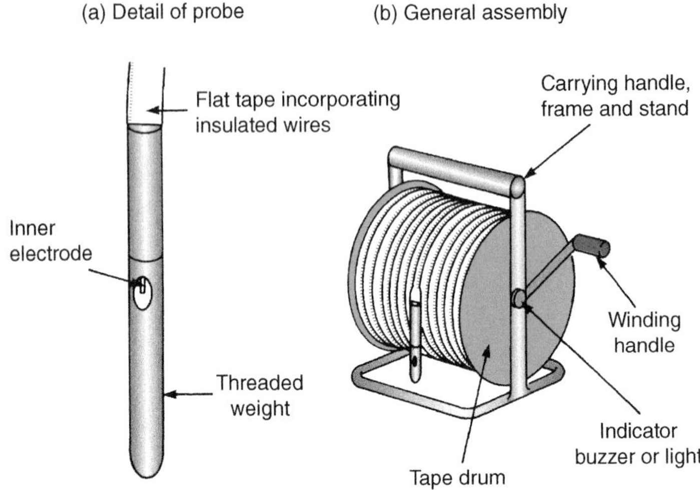

Figure 4.1 *Main features of a commercially available dipper.*

4.1.1 Using a dipper

Lower the probe down the well or borehole until it hits the water, when the buzzer will sound. When this happens, pull the tape back slowly out of the water until the signal stops. Repeat the exercise several times to enable you to 'feel' the water surface. It is conventional to take the point where the buzzer stops as being the water level. Use your fingers to mark the position on the tape against a fixed datum point such as the top of the casing, and then read off the level to the nearest centimetre. When using a dipper with a cable marked off in metres only, use a steel tape to measure the distance from your finger to the nearest metre mark *below* the place you are pinching, and *add* the two values together. Record the dipped level and time of measurement and make a note of the datum used. It is good practice to repeat the reading after you have written it down, as a check. It is a good idea to mark the datum point with coloured paint if you expect to take other readings in the same well.

Test the dipper each day before you go into the field by unscrewing the weighted end of the probe to expose the inner electrode, and then complete the circuit either by using the weight to short out the contacts or by immersing the end in water. A commonly used alternative is to wet the fingers with spittle to do the same job, but this is not good practice. If the buzzer does not work check the batteries, and should these be working inspect the cable for breaks. The first 30 m of cable gets most wear, and so the insulation can be rubbed off as the tape is lowered over sharp edges such as the top of the casing. Try to reduce this wear as much as possible by running the tape over softer materials such as a piece of wood or your hand. If the insulation is worn through to the wires they may break, thereby preventing the circuit from being completed. Do not be tempted to cut off the damaged section and reconnect the probe, otherwise you will always have to remember to deduct a constant length from your readings and errors will inevitably occur. Occasionally, the circuitry inside the drum may be faulty and will need checking and repair by an electronics expert. It is worth remembering that dippers will not work in groundwater with a low conductivity.

Great confusion can be caused in the field when the dipper works at the surface but fails down the borehole. In these circumstances, improvise a dipper that 'feels' the water surface. Do not add salt to the water in the borehole, even though this will increase the conductivity and enable the electric dipper to work. Such action will alter the groundwater chemistry and in some countries may be illegal. When making water level measurements in wells or boreholes that are used for water supplies ensure you adhere to both the hygiene and safety precautions given in Appendix I.

You may also need to measure the total depth of a well or borehole. There are commercially available total depth probes that consist of a weight (about 0.5 kg) attached to a length of steel wire cable marked off in metres. To measure the depth of the borehole, lower the probe until you feel the weight 'come off' the cable when it reaches the bottom of the well. Improvised total depth probes can be made using a length of string and a weight, but make sure that the string is sufficiently strong and the weight is not too bulky to fit easily in the borehole and is not made of material that could pollute the water. A weight tied to the probe of a dipper will also work, but the high water pressure at the base of the borehole may damage the seal on the dipper and so is not recommended.

4.1.2 Improvised dippers

If a commercial dipper is not available, it is possible to make your own, as shown in Figure 4.2. This example uses a door-bell cable and an electronic circuit. The cable should be weighted to make it hang straight, and allowed to stretch before markers are fixed at metre intervals. Take care that the markers cannot move or come off. The electronic circuit is needed to overcome the resistance of a long wire and consists of a transistor, a 30 kΩ resistor and an LED display light (or buzzer), and is powered by three 3V 'button' batteries. Make sure that the two bare ends of the cable cannot touch and the circuit will then only be completed when both electrodes are submerged. Test the circuit in a bucket of water before using in the field. An advantage of this type of home-made device is that the relatively thin wire will allow access through smaller holes at the wellhead than is possible with a commercial dipper and allow you to take readings on a borehole with difficult access.

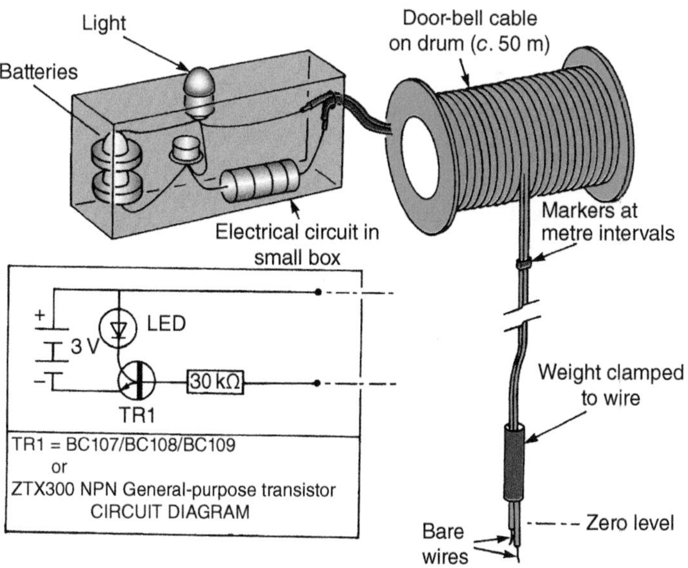

Figure 4.2 *A home-made dipper comprising a length of door-bell cable and simple electronic circuit made from components available from any electronics supplier.*

4.1.3 Measuring artesian heads

In boreholes where the water level is above ground level, head measurements must be made in a different way. Ideally, the borehole should be capped using a blank flange and the threaded dipping hole is fitted with either a pressure transducer, a pressure gauge or a transparent tube, as shown in Figure 4.3. In the example shown in Figure 4.3a a pressure transducer has been fitted to an artesian borehole that previously had only a pressure gauge installed. Standard plumbing fittings have been used to make a small chamber for the pressure transducer, which is held in place by a screw fitting that also grips the body of the transducer. Silicone sealant has been used to ensure that no water leaks occur. The type of device that combines the data logger with the transducer will avoid these potential leakage problems as the entire instrument is contained within the fitting and it is not necessary for a cable to pass through the casing. The original pressure gauge has been installed on a T-piece fixed to the side of the transducer chamber. Gate-valves have been installed so that the transducer can be removed without serious leakage occurring and the pressure gauge can be removed without interfering with the transducer record. Both the transducer and the pressure gauge have been installed at the same level so that the datum point for all water level measurements is the same. The pressure gauge was originally fitted immediately above the lower gate-valve, which means that the new datum is above the old one. It is essential that the record of water levels from the borehole is corrected so that a single datum point is used for the entire data set.

Choose a robust plastic tube and fix a metal screw fitting to one end. It is sometimes difficult to keep the tube absolutely vertical, but this will not matter, provided that you measure the *vertical* distance between the datum and water surface. More accurate readings are achieved by using a data logger system.

Although a pressure gauge has the advantage of being quicker and simpler to use than a tube, a standard pressure gauge is likely to have an accuracy no better than ±0.5 m, which may be suitable for regional monitoring but not a pumping test. A transparent tube can give an accuracy of ±0.5 cm and is much better for pumping tests. If the heads are more than 2 m above ground, however, practical considerations may dictate the use of a data logger or a pressure gauge. A long tube can be used if it is fixed to a convenient tree or telegraph pole. Remember, it does not matter how many curves there are in the tube, provided that you measure the *vertical* distance above the datum. Take care to ensure that the tube is not kinked so that it restricts the flow of water inside the tubing. It is important to ensure that you measure the true vertical height of the water level above a datum. If necessary, use a plumb line or spirit level to check it.

4.2 Continuous Water Level Recorders

Where a continuous record of groundwater levels is required, a recording device of some sort must be used. Although modern electronic equipment is generally used, it is a mistake to think that this is always the case. Groundwater-level records have been kept for many years and, for cost reasons, the original instruments are still likely to be in use as long as they are serviceable. In this section we examine the range of instruments used for recording groundwater levels.

Take a check reading with a dipper each time you visit the site either to download data from a logger or change a chart, making a reading both before and after you download the data and note the two values in your field book, not forgetting the date and times of the readings. Compare the value with the current reading of the instrument to assess the accuracy of your groundwater-level record.

4.2.1 Data logger systems

Most groundwater-level records are now made using a combination of a pressure transducer and a data logger (mini-computer). The system either comprises a pressure transducer installed in the borehole with a data logger housed at the wellhead or has the data logger inside the same case as the

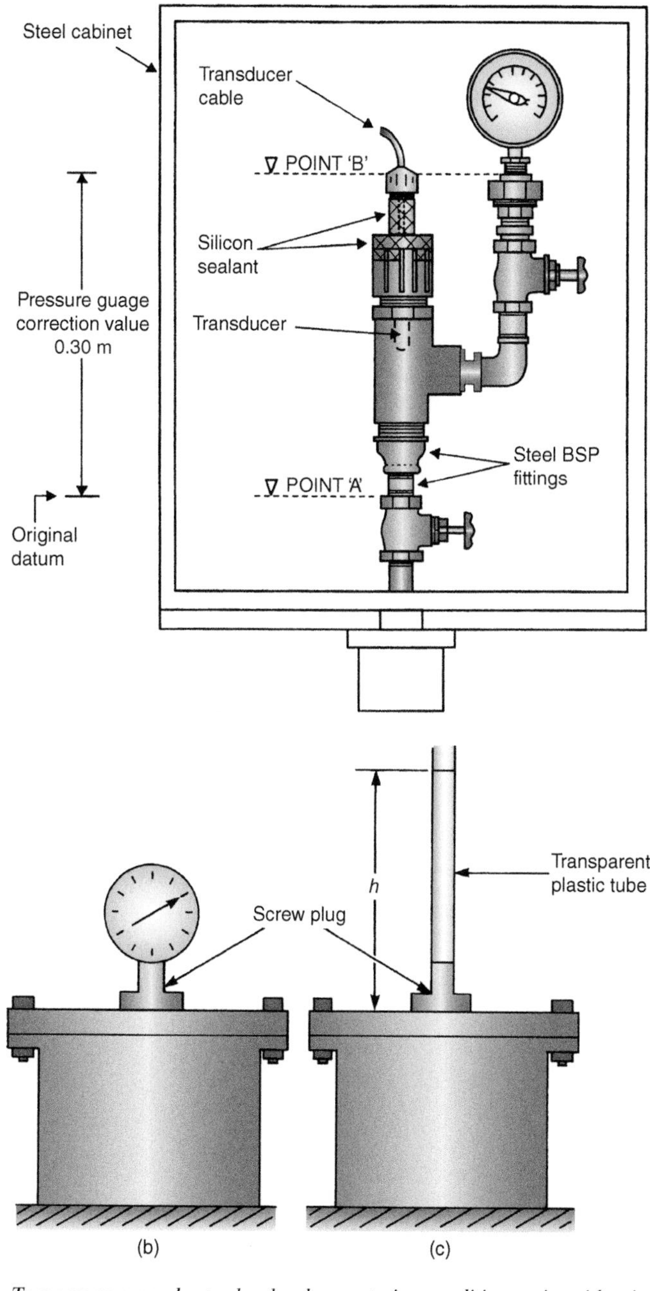

Figure 4.3 *To measure groundwater levels where artesian conditions exist, either install a small chamber to take a pressure transducer/data logging system (a) or use a pressure gauge which is calibrated in metres head of water (b), or, for greater accuracy, fix a transparent plastic tube to the borehole (c) and measure the water level as a head (h) above the datum point.*

pressure transducer, which is installed below the water level in the borehole. Data logger systems are much easier to use than any other form of water level recorder, and are invaluable for taking measurements during pumping tests or routine monitoring. Modern equipment allow data to be downloaded remotely onto a USB flash drive using wireless systems and avoid any interruption in data collection. There are a range of instruments available of different sizes both physically, the materials they are made from, and in terms of the data they can store.

The probe is installed in the borehole at sufficient depth below the water surface to ensure that it will not be exposed by changes in level. Fluctuations in the water level cause a corresponding change in the weight of water above the transducer, which varies the electrical resistance of the device. The data logger is programmable so that the time interval between readings can be pre-selected. When each measurement is taken, a small electrical current is passed through the transducer, with the variation in voltage being recorded by the data logger. This information is converted into water level values by the data logger.

All pressure transducers are pre-set in the factory to cover one of the standard response ranges. As the variation in electrical resistance over each pressure range is the same (usually 0.1%), the accuracy of the water level measurement is the greatest with the smallest range and decreases as the size of range increases. For example, a 1-bar transducer will have an accuracy of ±1 cm and a 10-bar transducer will have an accuracy of ±10 cm. Besides the accuracy (i.e. how well the transducer will reflect changes in water level expressed as a percentage of the range), you need to consider the resolution (or sensitivity), which is how much the water level will change before it causes a change in the transducer. Resolution is also expressed as a percentage of the water level range for the transducer. For example, a transducer may have a range of 0–25 m, an accuracy of 0.1% and a resolution of 0.025%. This will mean that each reading in the data set will be to ±25 mm and will be sensitive to water level changes in excess of 6.25 mm.

Table 4.1 lists the standard pressure transducer ranges. Choose a pressure transducer that will respond throughout the anticipated variation in water levels. For example, if you think that the maximum variation in water levels will be 12 m you will need a 1.5-bar transducer, which will respond up to 15.3 m. Selecting the correct transducer is important, especially for pumping tests where the water level range is likely to be large.

Table 4.1 *Standard ranges of pressure transducers.*

Pressure range (bars)	Equivalent water level range (m)
0.35	3.6
0.7	7.1
1.0	10.2
1.5	15.3
2.0	20.4
3.5	35.7
5.0	51.0
7.0	71.4
10.0	102.0
15.0	153.0
20.0	203.9
35.0	356.9

Values assume that 1 bar is equivalent to 10.197 m of head of water. After Cambertronics Ltd.

Installation of a data logger system is very simple and the manufacturer will provide a detailed step-by-step guide. Set the system up by connecting the data logger to your PC or laptop and use proprietary software to set to the required frequency for making each recording. You can do this in the office before you set off, in the back of your vehicle on site, or at the wellhead. Some systems use wireless connections, making life much easier. The recorded data are downloaded in the same way and again can be done in the field or by bringing the instrument back to the office.

You will then need to decide how far down the well you should hang it to make sure that the instrument is always below the water level and below the lowest level anticipated during a pumping test. For use in a pumping test you will need a range that will cope with the drawdown. Make sure that you install the transducer in the pumping well so that it cannot get sucked into the pump intake, either by installing it in a perforated dip-tube (which is the best option) or by hanging it at a depth below the bottom of the pump.

To install, just hang the transducer down the borehole on its cable and, depending on the system used, secure the data logger in place. Measure the water level with your dipper so that you can interpret the initial value of the record. Many modern systems have the logging done inside the probe, so they can just be hung inside the borehole. Some data loggers can be fitted inside the top of the borehole casing. Alternatively, the data logger will need to be housed in a small waterproof and vandal-resistant box located near the wellhead.

It is usually possible to hire electronic data monitoring equipment from the companies that sell it. They are often happy to install it for you and to download the data, presenting it to you on a computer disc or emailing it. Such services can be cost effective if you do not use this type of equipment on a frequent basis. Figure 4.4 shows the installation of a data logger system and how the data are retrieved on a routine basis. The photograph shows the installation of an instrument that has the data logger incorporated in the same housing as the pressure transducer and is installed in the borehole below the water level. The laptop computer is used to set the recording frequency and to download the data on site.

4.2.2 Float-activated chart recorders

Although now old fashioned, it is possible that you will come across float-activated chart recorders. Some of these instruments have been modified to have a battery-operated clock and perhaps even

Figure 4.4 *Installation of a data logging system requires the data logger and a laptop or PC to programme it, and this work can be done at the wellhead. (Courtesy of Van Essen Instruments Limited.)*

download the data onto a data logger and incremental encoder. There are two types, each sensing the water level by means of a float. Both have a revolving drum that holds a paper chart, and a clockwork mechanism for recording the time. In the first type (Figure 4.5), a perforated steel tape is attached to the float and passes over a toothed pulley with a counter-weight used to keep it taut. As water levels change, the float responds, moving the pulley, which is attached to a horizontal chart drum through a system of gears. A pen moves horizontally across the chart, driven at a steady rate by the clockwork mechanism to record the water level on the chart. The scale of the graph can be changed by selecting the appropriate gears to alter both the time scale and the vertical scale as required. In operation, the instrument is enclosed by a metal cover and usually is further protected by a lockable box. Beware of condensation inside the box, which may ruin paper charts. Also look out for insects, snakes, and other animals taking up residence; some of them may have a poisonous bite!

The alternative type of arrangement (see Figure 4.6) has a vertical chart drum that is driven round the clockwork mechanism. In this case, the pen is moved vertically by the rotation of a pulley wheel that is attached to the float in the same manner as before.

Some types use a waxed-paper chart and a sharp point to make a trace as the point scratches through the wax to reveal coloured paper underneath. This system has the advantage that it cannot run out of ink and the chart does not suffer from damp conditions. However, the chart surface is easily scratched, which may obscure the trace, and if the charts are stored in very warm conditions the wax can melt. Don't smile, it happened to me!

Horizontal drum recorders will continue to function when there are large changes in water level, such as during a pumping test. The disadvantage is that the chart must be changed within the specified time period or the record will be lost as the pen jams against the end of the drum. Vertical drum recorders will continue to function for as long as the clock continues to drive the chart drum, so data will not be lost if you are late in changing the chart. Both types of chart recorder may be used in river flow gauging stations.

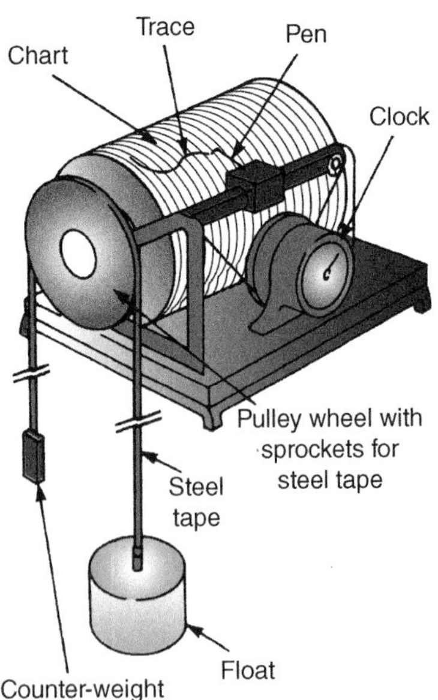

Figure 4.5 Horizontal drum recorders are arranged so that the drum revolves in response to changes in water level, while the pen is driven by a clock.

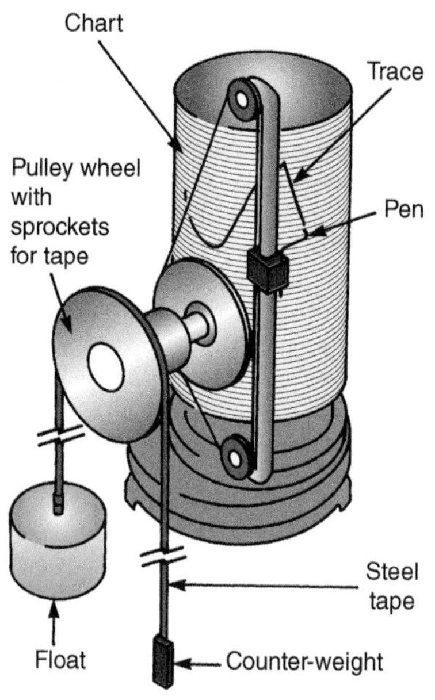

Figure 4.6 *Vertical drum recorders are arranged so that the pen moves in response to water level changes, while the drum is rotated by a clockwork mechanism.*

Chart recorders need regular maintenance visits to change the chart, wind the clock, oil the moving parts and fill the ink reservoir. Take several spare clocks so that you can replace any that have stopped or are slow. Check their accuracy in the office by comparing their time with a reliable clock, your mobile phone, or the time on your pc.

Some types of float recorder use a punch tape system to record the water levels instead of a chart and are powered by batteries. All these technologies are very old fashioned by today's standards; however, as the instruments were well made they may still be found on some established networks. The tapes are removed from site periodically and fed through a tape-reader in the office to produce a digital record. The limited availability of the tape-reading equipment means that these recorders are being replaced steadily with data logger systems.

4.3 Measuring Ground Levels and Locations

In order to interpret groundwater-level measurements and calculate groundwater-flow directions you need to relate the water levels to the same datum and be able to pinpoint the exact position of each observation point. Elevations are measured as a height above a datum that usually approximates to sea level. Locations may be defined as latitude and longitude or more commonly by a locally used map referencing system such as the UK Ordnance Survey national grid system.

4.3.1 Using a surveyor's level

The most accurate ground-level measurements are obtained by using a surveyor's level. That is essentially a telescope mounted on an adjustable base that enables the line of sight through the telescope to be maintained in a horizontal plane. Ground levels are measured against this horizontal line, using a levelling staff – a sort of long telescopic ruler. The elevation of each location is measured against the starting point, either where the elevation above sea level is already known or it is a 'temporary site datum point' where you have assumed an elevation using local maps. Levelling using this type of equipment is a straightforward procedure, and with care you will be able to obtain results to within a centimetre or two, although professional surveyors are much more accurate.

Figure 4.7 illustrates how this equipment is used, and a more detailed description can be found in textbooks on surveying (e.g. Bannister *et al.*, 1998). A surveyor's level must be set in a horizontal plane at each station where it is used. Set up the tripod and make the base-plate as horizontal as possible. Then adjust it using the three levelling screws, using the small bubble indicator to show you when it is absolutely horizontal (Figure 4.7a). This procedure will be repeated at every station where you set up the level. Start to level from a location where you already know the elevation. Ask a helper to set the levelling staff on this point so you can focus the instrument's telescope onto it. The staff must be held vertically using a bubble-gauge on one side as a guide. The helper must hold the staff still and ensure that his or her fingers do not obscure the front of the staff. This is easier if the staff is only extended by the minimum amount required at each location. Look through the eyepiece and focus the telescope on the staff so that you can both read the numbered markings and see the crosshairs in the telescope. Read off the height value from the staff that is covered by the central horizontal cross hairs. The staff is divided into 1-cm gradations, with numbers marking off every 1 m and 0.1 m. Carefully count up to the mark covered by the cross hairs from the closest value below it to work out the height above ground level at that point. Once you have read the value, write it down and then check this reading by looking through the telescope again. Double-check each reading after you have written it in your field book. The level remains in the same position, while the staff is now moved to the next point where you want to know the level. The same procedure is followed to use the staff to measure the vertical distance from the line of sight to the ground at the second point.

The relative height between the two points is the difference between the two values read off the levelling staff, as shown in Figure 4.7b. In most cases, you will have to move both the staff and the level several times to move from a point of known elevation to the place where you need to know the height. The next move would be to keep the staff in the same position and set up the level at a

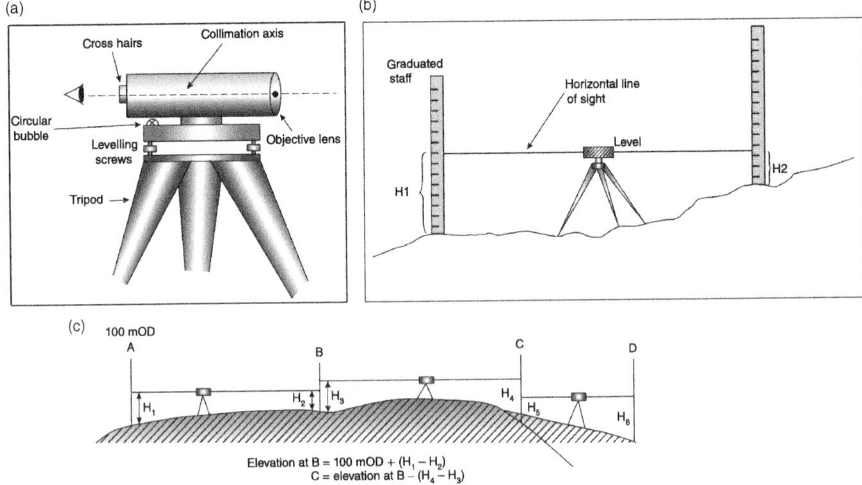

Figure 4.7 *A surveyor's level (a) and how it is used (b and c).*

new location towards the end point. Make sure that your helper does not move the staff off the point where the first level was taken while you are moving. Once you have made the base-plate level, read the value on the staff to measure the height of your new line of sight using the same procedure. The staff is then moved and the process repeated as many times as necessary, with the level and staff 'leapfrogging' until your helper is able to place the staff level on the position where you need to know the elevation.

To calculate the level, simply add (or subtract) these values to find the difference between the two end points (Figure 4.7c). The surveying techniques that you require are not difficult, can be easily learned and are soon acquired with practice. You do not have to buy the equipment as both the surveyor's level and levelling staff can be hired by the day from survey equipment suppliers, although if you use them only a few times, buying a set can be better value.

4.3.2 GPS instruments

The global positioning system (GPS) was developed for the US government for military purposes, although during the 1980s the system was made available for anyone to use without charge. There are 24 GPS satellites that circle the earth twice a day in very precise orbits, transmitting information signals. GPS receivers use triangulation to calculate the user's exact location by comparing the time a signal was transmitted from several satellites with the time when it was received. The time difference tells the GPS receiver how far away each satellite is from the receiver and allows the user's position to be calculated.

A GPS receiver must be locked on to the signal from at least three satellites to calculate a 2D position (latitude and longitude or grid reference) and track movement, and signals from at least four satellites are needed to determine the user's 3D position (latitude, longitude, and altitude). GPS readings are extremely accurate to within 15 m on average, which is good for location but poor for elevation. Some modern systems can be accurate to less than 3 m on average, although they used WAAS (Wide Area Augmentation System) and are currently only available in North America. When you use your instrument make sure that you read the instructions supplied by the manufacturer and that you are out in the open so that the satellite signals can be received. As always, it is best to try it at home or in the office first.

4.3.3 Estimating elevations from topographical maps

Ground levels can be estimated from topographic maps to an accuracy of less than half the contour interval. Most British Ordnance Survey 1:25,000 scale maps have an interval of 5 m, so you should expect to be able to estimate levels to ±2.5 m. This may be adequate for a general idea of groundwater flow directions in hilly areas, but is unlikely to be good enough in areas of low relief.

4.3.4 Map grid reference

The topographic maps for most countries have a grid reference system or sometimes use latitude and longitude to define locations. The map reference should be given as accurately as possible to identify the borehole ideally to within 10 m on the ground and certainly never more than 100 m.

The National Grid System used on UK Ordnance Survey maps comprises two sets of lines: one that runs approximately north–south (confusingly called Eastings because they are numbered in a west–east direction), and the other that runs east–west (Northings). The grid system forms a series of squares. Squares with sides 100 km long are identified by a pair of letters (e.g. TQ). The Eastings and Northings within each of these squares are numbered in kilometres from 00 to 99. Each kilometre square can then be divided with lines at 100-, 10- and 1-m intervals. Grid references are written with the letters first and then the numbers of the Eastings followed by a space and then those for the Northings. In practice, grid references are often quoted to six figures (e.g. TQ 435 867) with a precision of 100 m, or to eight figures (e.g. SJ 6594 9545) with a precision of 10 m. Guidance on the grid reference system with an example is given in the key section on all the Ordnance Survey maps. There are many other map grid systems used round the world, all of which use the same basic principles as described here.

4.4 Tool-Box

A wide assortment of tools is required for a variety of jobs that you are likely to face when working in the field. Use Table 4.2 as a guide to put together a tool-box to take in the field and avoid leaving an essential tool behind.

4.4.1 Removing well covers

Access to many wells is through a manhole cover. These are often made of cast iron and range widely in size and weight. Some have lifting handles incorporated into them, while others require the use of special lifting keys. Before removing a cover, clean off any soil, vegetation or other debris that may inhibit you lifting the cover easily or that could fall into the well once the cover has been removed. Use a hand brush to remove the dirt that has slipped down the small gap between the cover and the frame. Take care when lifting not to put either your hands or your feet under the cover and risk fingers or toes being broken if the cover slips. It is tempting to use a chisel or the edge of the spade as

Table 4.2 *Checklist for tools and equipment.*

18-inch adjustable pipe wrenches
Small spanners or socket set
4-lb hammer and claw hammer
Assorted chisels and crowbar
Manhole-lifting keys
Hacksaw, screwdrivers (large and small), pliers, allen keys
Wood saw, knife, brace-and-bit
Steel tape, builder's spirit-level
Spade, bricklayer's trowel, bucket, hand-brush, wire brush
Electrician's tape, assorted pieces of electrical wire and cable
Assorted nails and screws, 'super' glue, wood adhesive, gaffer tape (or duct tape), bolts, nuts, washers, releasing agent
Other equipment
Dipper, with spare batteries, depth sampler, bottles, funnel, pH meter specific ion probes, electronic thermometer (also see Table 7.4)
Total depth probe, stopwatch, measuring jug, pressure gauge, transparent tube
Torch (non-sparking type, in case of exploding gases)
GPS instrument
Surveyor's level, staff, measuring tape
Laptop computer
Safety equipment
Safety harness with rope, safety helmet, safety shoes/boots, high-visibility jacket, goggles, rubber gloves, heavy-duty 'rigger' gloves
Gas detector, miner's safety lamp, first-aid kit, mobile (cell) phone

a lever to prise up covers where the lifting handle is missing or when your lifting keys do not fit. This approach is not recommended as the cover can easily slip, causing injury or damage. Caution should also be exercised when attempting to lift heavy covers. Do not try this on your own; some of these covers are over 50 kg and could easily cause back sprains. Boreholes may have bolt flange covers fitted to the top of the casing (see Figure 4.10 in Section 4.6.2). Long-handled adjustable spanners or pipe wrenches, and perhaps a releasing agent, may be needed to remove them, but use a cloth or absorbent paper to make sure that none of the releasing agent gets into the water supply. Sometimes the borehole is situated in a small chamber below ground level that restricts access and prevents the use of long-handled spanners. A socket set will come in handy here. Avoid using the wrong tools as this is likely to damage the bolt heads or make you slip and injure yourself.

Many hydrogeological field trips require improvisation to be able to take the measurements you need, so include a few items in your tool-kit that may prove handy. Electrician's tape used for insulating wiring is made from vinyl and stretches well and has a large number of potential uses in the field. Many water-well drillers I know use it to temporarily hold cables, pipes, and so on, and I have used it to hold instruments in place at a well head. Similarly, gaffer tape and duct tape that are both stronger than electrician's tape have many potential uses in the field. I have used these types of tape to make a temporary repair on a pipe during a pumping test, for example. Both are fabric-based with strong adhesive properties. A word of warning, however; if you are taking samples for chemical testing that includes organic compounds, beware that the tape and/or the adhesive may affect your results.

Other items to carry are polythene bags that can be used for soil and rock samples but also in improvised flow measurements where there is no room to insert a jug or bucket.

4.5 Well Catalogue

Make the best use of the water level data and other information from the wells and boreholes in your study area by compiling a 'well catalogue', that is a database listing all known water wells and boreholes in an area. Springs should be included, as these also provide information on groundwater levels and represent groundwater discharges. Even abandoned wells, which may have been backfilled or covered over and lost, should be included if you have no information about them other than the location. Each entry in the well catalogue should be identified with a unique number that will give easy access to the details for each site. Table 4.3 provides a list of the headings that you could use as the basis for developing your own database. This information can easily be kept in a computer database such as Microsoft Access or in a card index system if computer systems are not available.

A sketch plan is invaluable when you visit a site after a long time and the vegetation has grown up. Measurements from the borehole to fixed points such as the corner of a building or a fence post can make all the difference in finding the borehole again. Include a diagram of the head-works arrangements, giving dimensions and the reference datum used for water level measurements. This helps when deciding which wells and boreholes are accessible and suitable for observation purposes and water sampling. A photograph or two may also help taken with your digital camera or mobile phone. The example in Figure 4.8 shows a sketch of a site where boreholes could be easily confused, and provides enough information for each one to be identified.

4.6 Field Surveys for Wells, Boreholes and Springs

Once you have obtained all the available information from existing records you will have to go out into the field to check details, measure water levels, and search for additional wells, springs, and ponds. If your project is for a new abstraction in England or Wales, the Environment Agency or Natural Resources Wales are likely to require you to carry out such a survey to identify all 'water features' within a defined area.

The best way to find wells is to ask local people. The majority of villages in developed countries have a mains water supply, but outlying farms and houses will probably have their own source of

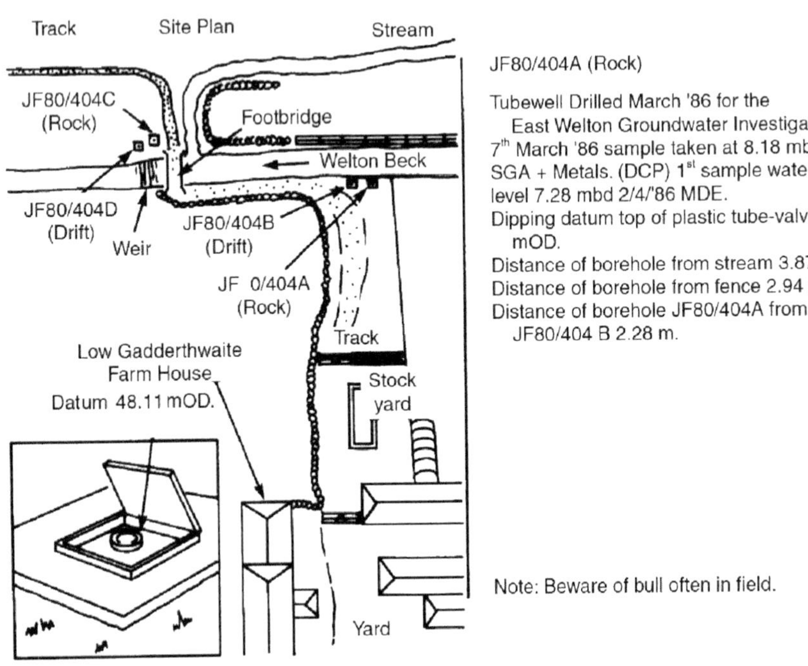

Figure 4.8 *Make a sketch plan to help locate the borehole in future and keep it with records for the site. Access difficulties and hazards such as bulls and dogs (or even a talkative farmer) can also be noted.*

water. The only way to find out is to knock on the door and ask. This requires diplomacy if the enquiry is to be fruitful. Be prepared to spend time explaining who you are and what you are doing, but be prepared to be brief so as not to waste the time of busy people. When undertaking this sort of survey, take a large-scale (e.g. 1:10,000, or 1:25,000) map along with you that shows field boundaries. Ask each farmer to describe the boundaries of his property and then identify them on the map and point out the position of all his wells, boreholes and springs. Do not expect everyone to be able to read a map, so visit each site with the farmer if necessary. He should also be able to tell you the name of his neighbours, so that your next visit will not be completely uninformed. In this way you will build up a picture of all the local sources, and any fields missed from your survey should be obvious.

Note the position of ponds and boggy areas as these may indicate areas of groundwater discharge, and stay on the lookout for signs of wells and boreholes. A hand pump is an obvious sign, but it could be misleading as they are not always on top of wells. They are also used to pump water from underground rainwater collection tanks, for example, so look under the manhole cover and check it out. They may even be used as a garden ornament and have nothing at all to do with water! Wind-pumps are more easily seen, even when they are broken. They are generally sited over boreholes or wells and unlikely to be over a rainwater collection tank.

Sometimes tall 'thin' buildings may contain a borehole. The height is needed to house lifting gear to pull out submersible borehole pumps and sometimes to house a water tank on the roof to provide sufficient head to drive water round a distribution system. Lifting tackle, in the form of a sturdy frame, may be permanently installed over boreholes that are not in a building. Keep your eyes open. Eventually you will develop an instinct for finding wells and boreholes and will even be able to spot them from your car as you drive by. When you visit each well, borehole or spring, obtain as much information as possible for your records. Take as many measurements as are practicable, rather than

Table 4.3 Headings for well catalogue.

Heading	Comment
Name of site	Use an obvious name and make sure that it is unique to the well in question. Identify multiple wells separately.
Well catalogue no.	This is the identification number relating to your well catalogue.
Abstraction licence no.	Only applies where water abstraction is controlled by law.
Geological Survey no.	National geological survey organisations may have a well record numbering system making cross-referencing easy.
Map reference	This is the unique identifier for the map system used in the country where you are working. See Section 4.3.4.
Status	Is the borehole used for water supply, disused, or a purpose-drilled monitoring borehole, and so on?
Aquifer	Record the stratigraphic name of each aquifer here so that the database can be searched by aquifer name.
Depth	The total depth of the borehole.
Diameter	Record all drilled diameters in the borehole.
Casing	Note depth of top and bottom of each length of casing used, its diameter and also the material.
Construction date	Note the month and year when the borehole was finished.
Groundwater levels	Note all water strikes during construction and the final rest water level after the borehole was finished. Include any more recent readings including information on whether the borehole was being pumped.
Datum	Relates to ground elevation above sea level or map datum.
Notes	Record any useful information. Include information on access or other abstractions nearby.
Site plan	A sketch plan noting the borehole location from fixed points, such as the corner of a building or fence posts, to help find the borehole in the future. Use at least two fixed points.

relying on the owner's knowledge. It is not unknown for owners to exaggerate about well yields and the reliability of their springs or even simply to be mistaken.

4.6.1 Locating springs

There are tell-tale signs to look for when locating springs. Your geologist's eye for the country will help to identify spring lines, such as the example in Figure 4.9. Vegetation is a great help; rushes and sedges that grow in wet places are often a darker or more lush green than the grass covering the rest of the field or the hillside. Clumps of these plants can often be seen at intervals, thereby clearly taking your eye to the spring line. Look out for spring collection chambers and storage tanks that are part of a water supply. Sometimes a hydraulic ram is used to pump the water. These make a characteristic 'clunking' noise at regular intervals that will allow you to 'home in' on both hydraulic ram and spring.

Not all water that issues from the ground is spring water. It may be water discharging from a land drainage pipe or culvert and consists largely of surface water. Other alternatives include leaking water supply pipes and sewers which you may be able to identify from their chemistry. All true springs consist of groundwater and may be identified by their chemical or physical characteristics. Springs will have a similar chemical composition to local groundwater, and a comparison of chemical analyses will help separate springs from land drains or leaking pipes. Leaking treated water is likely to have different chemical characteristics from the groundwater and may also have

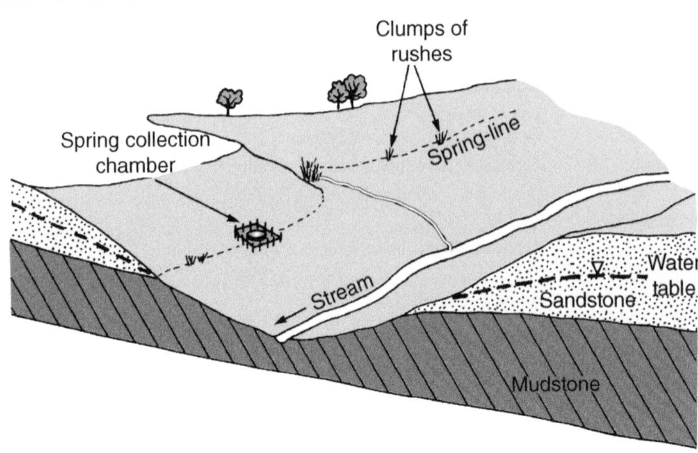

Figure 4.9 During your walkover survey you can spot a spring line by observing the vegetation and other signs. This spring line follows the contact between a sandstone and underlying mudstone and is marked by clumps of rushes or sedges, the start of a minor tributary to the main stream and a spring collection chamber. Most chambers look like a masonry or concrete box, partially buried and usually fenced round (see Figure 10.2).

traces left from chlorine-based disinfection treatment (see Section 9.3.2). Sewage in groundwater can often be identified by the presence of faecal bacteria and increased concentrations of some chemical species such as chloride, ammonium, nitrite, and nitrate.

Except in geothermal areas such as Iceland, New Zealand, the Dead Sea, and other sites near active or former plate margins, groundwater temperatures down to about 100 m depth remain relatively constant at around 9–15°C throughout the year, approximating to the local average air temperature. As a result, groundwater seepages and springs are easy to identify during the winter, because they are unlikely to freeze and will support plant growth at a time when there is none elsewhere. Conversely, during the summer, spring waters will appear cold. Use a thermometer to compare water temperatures. For best results take the readings during the afternoon of warm or hot days when surface waters will have had the opportunity to warm up. This technique can also be used to locate areas along a stream or river where there is a significant groundwater discharge. Temperature readings should be taken with a thermometer at a frequent interval (such as 10 m) along the stream, and the results plotted as a longitudinal profile of temperature. Groundwater discharge areas will stand out, provided that the survey has been carried out on a day when sufficient temperature contrasts exist. Some remote sensing techniques that employ infrared cameras can also be used to pick out temperature differences in a similar way.

4.6.2 Problems in measuring water levels

Access to large-diameter wells to measure the water level and total depth is usually easy. Beware of the atmosphere in wells, especially if they are capped with a manhole cover, and do not enter until you have taken measurements and are satisfied that the atmosphere is safe. The air may be depleted in oxygen or sometimes even explosive or have a high radon content. Read the advice contained in Appendix I and be careful! Remember, your safety is *your* responsibility.

Figure 4.10 shows three examples of different arrangements for borehole head-works. Purpose-drilled observation boreholes (Figure 4.10a) should be fitted with a welded flange to which a blank flange is bolted. A dipping plug of 25–50 mm diameter installed in the centre of this flange will

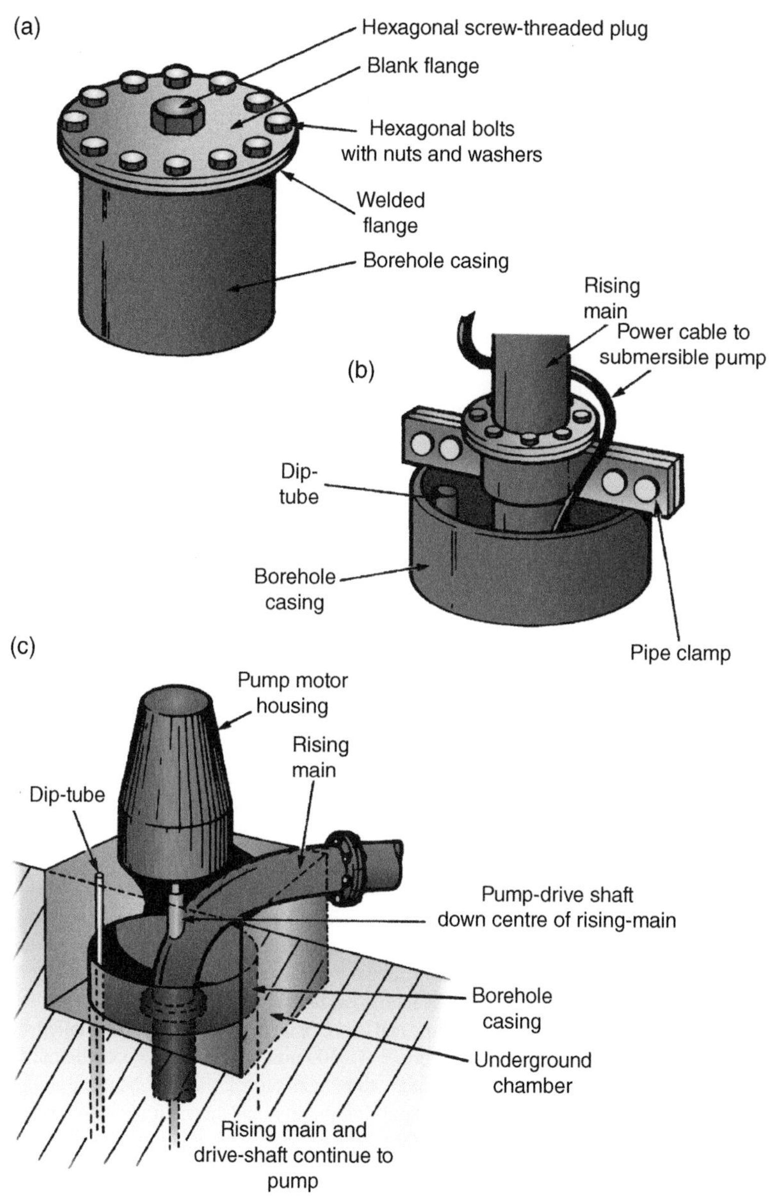

Figure 4.10 *A variety of head-works are found on boreholes and this diagram shows three examples.*

make water level measurement easy. Small-diameter boreholes should be capped with a screw-threaded top straight on to the casing.

Many abstraction boreholes have electrical submersible pumps hung on lengths of flanged steel rising main (Figure 4.10b). The top of the borehole may be at ground level or in a chamber. Commonly, the rising main is hung on a pipe clamp, which is fastened round the main just below one of the flanges. In many cases, the diameter of the borehole is only 50–100 mm larger than the flanges, giving amore restricted access to the borehole than indicated in the diagram. The installation on some small-diameter boreholes (200 mm or less) includes an extra-large flange on the rising main in the position equivalent to the pipe clamp. This allows the pump to be hung on the flange, rather than a clamp, but restricts access for water level measurements. Sometimes one of the bolts is left out, to permit access through the bolt-hole. It is strongly recommended that a dip-tube is installed in all abstraction boreholes to avoid the risk of dippers becoming snagged on the pipework or the power cable.

Some abstraction boreholes contain shaft-driven turbine pumps driven by a surface motor (Figure 4.10c). These motors have to be installed immediately over the borehole, thereby restricting access for water level measurement. Such installations commonly have a small underground chamber as part of the head-works, as shown in the diagram. When dipping abstraction boreholes, look for an access hole in the 'capping' flange on the rising main. If you are lucky there will be a 'dip-tube'; these should be 20–50 mm in diameter and open-ended, with the lower 5 m or so being perforated to increase the contact between the tube and the water in the borehole. Dip-tubes also prevent false readings that are often caused by water entering the borehole above the pumping-water level.

Without a dip-tube there is a high chance that the dipper will get snagged in the borehole. Boreholes are seldom absolutely vertical and pumps are rarely hung centrally, and the electricity cable to the pump is often loosely attached to the rising main to form loops that can catch your dipper. Space is more limited when the pumps are hung on flanged pipe. Even with unrestricted access at the wellhead, the chances are that whichever side of the rising main you choose to lower your dipper it is likely to become entangled. This is made worse by the Earth's rotation, which will cause the dipper to swing like a pendulum, especially if the groundwater level is more than about 30 m down.

If your dipper becomes stuck, do not keep pulling! Give it a waggle and try to lower it further down the hole before attempting to move it sideways round the borehole. If all else fails and it looks as though the dipper is stuck fast, you may as well disconnect the tape and leave it at the wellhead to be recovered when the pump is next pulled out for maintenance.

Some older large-diameter boreholes have a water level indicator installed that works on air pressure, as shown in Figure 4.11. The pressure vessel is charged with compressed air, using a foot-pump, and the pressure measured on a gauge calibrated to read as a depth below floor level. Air flows from the pressure vessel through the air tube down the borehole. The pressure-gauge needle will gradually rise as pressure in the system increases, but settles to a steady reading once the air pressure is equal to the head of water above the end of the open tube. When newly installed, this type of device probably gives a reading to an accuracy of ±300 mm, which is not good enough for most hydrogeological purposes. Even for this accuracy to be maintained, the air tube must not leak, and each time that it is removed from the borehole during pump maintenance it must be replaced with its bottom at exactly the same level. This rarely happens and, consequently, these devices must always be regarded as suspect. It is best to avoid using them altogether and never, ever use them for a pumping test! Having no data is always better than wrong data!

4.7 Interpretation of Abstraction Borehole Water Levels

Information on the borehole construction details or at least its depth is needed to interpret groundwater-level data and understand how it relates to groundwater flow in the aquifer. A record of groundwater levels will help you to decide whether a particular aquifer is confined or unconfined (see Figure 1.2). Figure 4.12 shows three typical situations and how the water levels relate to the groundwater

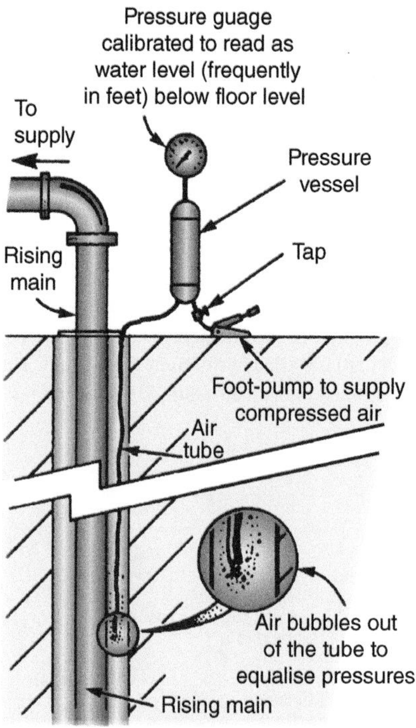

Figure 4.11 *Air pressure is occasionally used to measure water levels in abstraction boreholes, especially where access is restricted, such as the case in Figure 4.10c. These instruments frequently give erroneous readings, and it is best to avoid using them.*

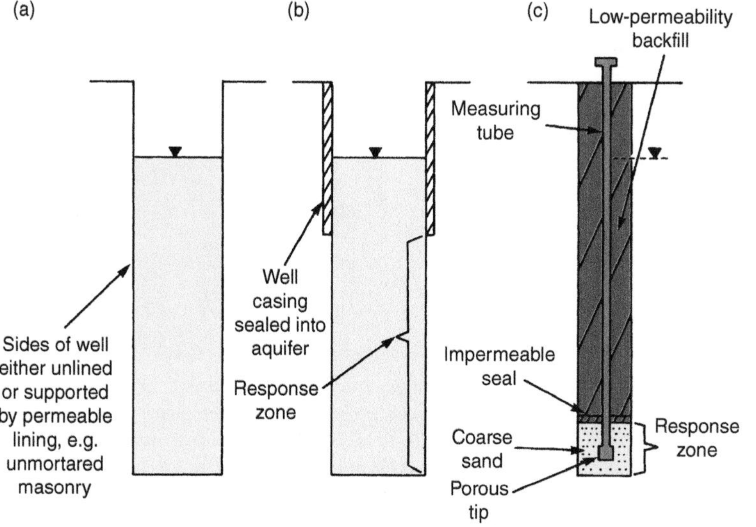

Figure 4.12 *Relationships between the well or borehole construction and groundwater levels.*

conditions. Groundwater levels in an uncased well or borehole (Figure 4.12a) will represent the local *phreatic surface* (i.e. the water table). In a cased borehole (Figure 4.12b) the water level represents the average hydrostatic pressure (i.e. *pressure head*) in the uncased section of the borehole. Depending on the local hydrogeology, this may be either a phreatic surface or, if the aquifer is confined, a *piezometric surface*. The water level in the piezometer (Figure 4.12c) represents the hydrostatic pressure (i.e. pressure head) at the level of the response zone. Except where the piezometer is just below the water table, this water level will not reflect the phreatic surface. If the borehole penetrates an unconfined aquifer the water level represents the water table. On the other hand, the water level in boreholes penetrating confined aquifers will be above the top of the aquifer and represent the piezometric surface.

Where small-diameter pipes are used to measure groundwater heads they are usually referred to as *piezometers* and are often installed round landfill sites and areas of contaminated land to examine groundwater conditions or as part of a geotechnical investigation for civil-engineering construction. Sometimes several piezometers are installed at different levels in the same backfilled borehole and are referred to as a 'nest' of piezometers, with an example shown in Figure 4.13. The small-diameter

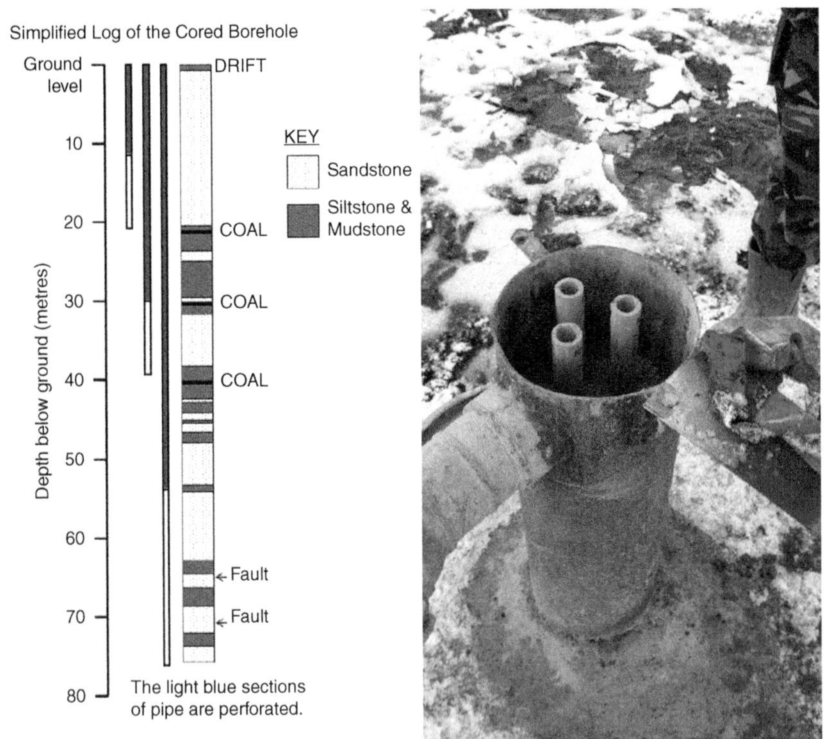

Figure 4.13 *This borehole was drilled to a depth of 75 m near Morpeth in Northumberland in the Stainmore Formation at the top of the Namurian Group. Three piezometer pipes were added with response zones at 75–54 m, 38–29 m, and 20–11 m. Each piezometer was surrounded by gravel and the rest of the borehole was filled with a sand/bentonite mixture. The left-hand drawing shows the relationship between the piezometer and geology, with different water levels in each piezometer. The water level in the highest piezometer is above that in the next one down; confusingly, that in the deepest one is above that in the highest one. The right-hand photograph shows the surface arrangement.*

pipe in Figure 4.12c has a perforated tip installed at the bottom of the borehole surrounded by filter sand and sealed in place with high-grade clay (e.g. bentonite). The remainder of the borehole is filled with a low-permeability material. Each piezometer reflects the value of the pressure head at that point in the aquifer and the differences reflect the local vertical hydraulic gradient.

Figure 4.14 shows records from a borehole located in the Sherwood Sandstone near Ormskirk, Lancashire, where the upper and lower piezometers are separated by 73 m difference in depth and show a constant head difference of 3–4 m in the water level record that is maintained throughout the period of the record. Seasonal fluctuations reflecting variations in annual recharge are clearly visible in the latter four years because of the increased frequency of measurement. The downwards gradient indicates that the location is a recharge area. The 3D distribution of head within an aquifer means that the water level in piezometers at different depths will not be the same.

Hubbert (1940) produced a ground-breaking demonstration of how groundwater flows through aquifer systems based on Darcy's law. In essence, he showed that both the depth of a borehole and its location within an aquifer have a significant influence on the water level within it. Figure 4.15 shows how these factors impact on groundwater levels. A further example is provided in Case History 1, and the significance of the 3D distribution of head that drives groundwater flow is discussed further in Section 4.11.

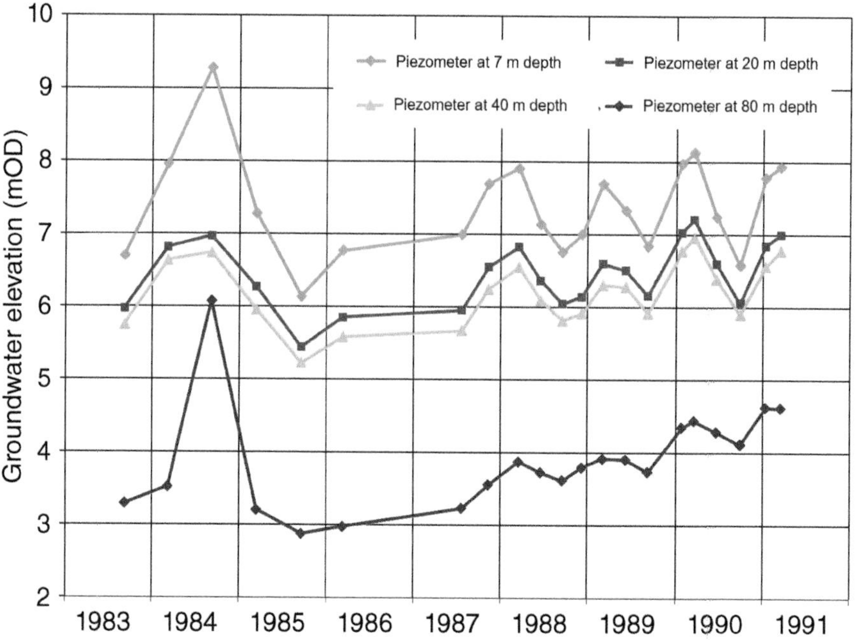

Figure 4.14 *This series of hydrographs shows the record of groundwater levels measured in a piezometer nest installed in a borehole in the Sherwood Sandstone aquifer in west Lancashire, England, 1983–1991. The downwards vertical gradient indicates that the location is a discharge area. (From Brassington (1992) by permission of CIWEM.)*

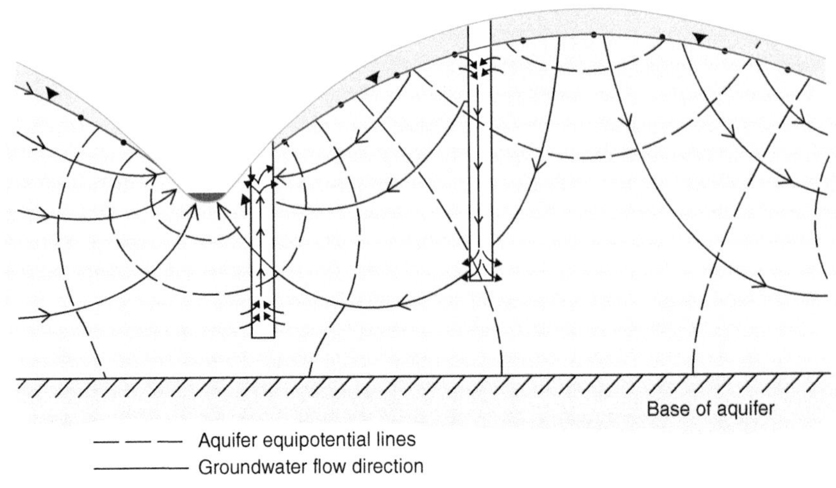

Figure 4.15 *How water flows through an idealised aquifer from recharge areas to discharge areas in valley bottoms. Flow lines are shown as solid lines, while equipotential lines are hatched. The effect of this flow system on water levels in wells gives decreasing elevations in recharge areas with depth and increasing elevations in discharge areas. Note that the water level in adjacent boreholes is not the same if the depths are different. (From Hubbert (1940) by permission of University of Chicago Press.)*

Case History 1 – Interpretation of Groundwater Strikes

Three site investigation boreholes were drilled using shell and auger methods (see Section 9.1.4) to depths of about 17 m into an alluvial sequence infilling a buried glacial valley at Bangor, North Wales. The strata penetrated were described from the cuttings obtained from each borehole and are summarised in Box Figure 1.1. It can be seen that two boreholes encountered the rockhead (Silurian mudstone), whereas the deepest borehole (No. 2) is located closer to the centre line of the buried valley and did not prove the rockhead.

Temporary casing was driven during drilling to support the sides, thereby limiting groundwater inflow to the bottom of the borehole. Each time that groundwater was encountered, work was suspended for 20 minutes to allow the water level to recover, following standard practice. There were three groundwater strikes in each borehole and in every case the water level rose up the borehole (see Box Figure 1.1). The rest-water level for each first strike lies above that for the second strike, implying a potential downward hydraulic gradient. These rest-water levels approximate to a culverted watercourse that runs across the site. The deepest water strike in each borehole resulted in a rapid rise above the levels relating to the other two strikes and just below the ground level.

This relationship implies a potential upward flow from the deeper part of the geological sequence, indicating that there may be two groundwater systems.

Groundwater chemistry was used to test this idea. Water samples were taken of the deep and shallow groundwater and analysed for the major ionic content, with the results shown in Box Table 1.1. The total amount of dissolved minerals in the deep sample is only some 70% of the strength of the shallow sample, as shown by the conductivity values.

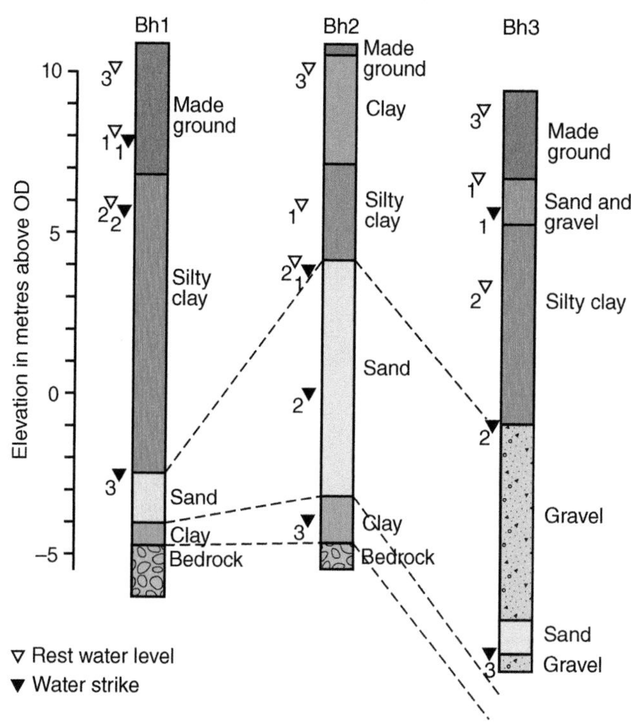

Box Figure 1.1 *Three site investigations drilled in Bangor, North Wales, encountered a sequence of alluvial deposits that infill a glacial rockhead valley. Three water strikes were encountered in each borehole, with the water level rising inside the casing. In each case, after the deepest strike, the water level rose 15 m or more to rest just below ground level. (Data by courtesy of Shepherd Gilmour Environment Limited.)*

Box Table 1.1 *Groundwater chemistry at Bangor, North Wales.*

Determinand	Deep	Shallow
pH	8.20	8.22
Conductivity (μS/cm)	466	657
Calcium (Ca)	49.86	75.41
Magnesium (Mg)	13.74	18.25
Sodium (Na)	14.4	24.8
Potassium (K)	2.4	3.6
Chloride (Cl)	26	42
Sulphate (SO_4)	51	98
Alkalinity (HCO_3)	140	180
Nitrate (NO_3)	<0.3	0.3
Nitrite (NO_2)	0.06	0.05

Chemical concentrations are expressed in milligrams per litre.

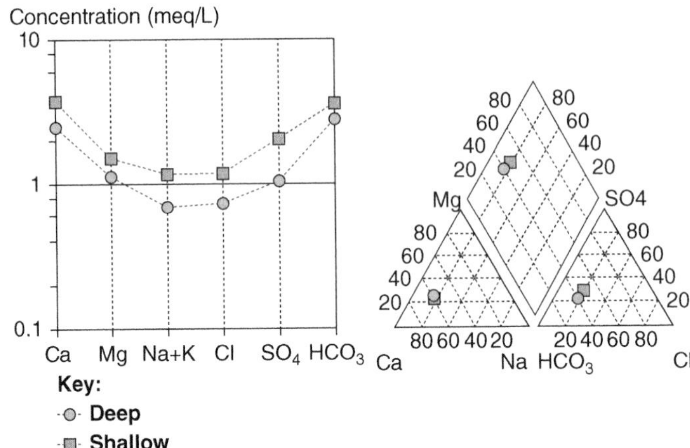

Box Figure 1.2 *Differences in chemistry of water samples taken from the shallow and deep groundwater systems encountered by the site investigation boreholes shown in Box Figure 1.1, illustrated by Schoeller and Piper graphs. (Data by courtesy of Shepherd Gilmour Environment Limited.)*

Schoeller and Piper plots (see Section 7.5) shown in Box Figure 1.2 were used to compare the data. The two key differences are (1) the overall concentrations of all the ionic species, indicated by the separate lines on the Schoeller graph, and (2) the dissimilarity in the ratio between the anionic content, indicated by these lines not being parallel and the differences between the positions of the points on the two triangular plots in the Piper graph. The deep sample data fall below the shallow sample values on the Schoeller graph, confirming that the deeper groundwater has a lower concentration of dissolved minerals. Although the two graphs follow a similar pattern, the lines are not quite parallel, indicating small differences in the proportions of some constituents and supporting the idea of a separate origin for the two waters. The differences between the two waters are more easily seen on the Piper diagram. The left-hand trilinear graph shows that the cationic make-up of the two waters has the same proportions. The right-hand trilinear graph, however, shows that the waters are different, with the deep sample having proportionately more alkalinity (HCO_3) and less sulphate (SO_4). From Box Table 1.1 it can also be seen that the shallow sample contains both nitrate and nitrite, whereas the deep sample contains no nitrate (NO_3), although nitrite (NO_2) is present. This indicates that the deep groundwater is depleted of oxygen and contrasts with the shallow system.

The evidence from the groundwater chemistry supports the idea that there are two separate groundwater systems. The groundwater-level evidence indicates that a shallow water table drains towards a (now culverted) watercourse, and a deep confined system is flowing towards the coast about 1 km away. The hydraulic head driving this deep flow to the sea has a strong upward component, as would be expected in a groundwater discharge area (see Section 4.7). It is concluded that the two systems are kept separate by a large contrast between the horizontal and vertical values of hydraulic conductivity that are expected, given the layered nature of the sediments.

When recording a water level in a supply borehole, note whether it is being pumped or is at rest and how long since the pump was turned on or off. Figure 4.16 shows the components of drawdown in an abstraction borehole and illustrates why it can take a very long time for natural groundwater levels to be restored after the pump is turned off. As water is pumped from the borehole (Figure 4.16a) the water level falls rapidly at first and then gradually slows down until a stable

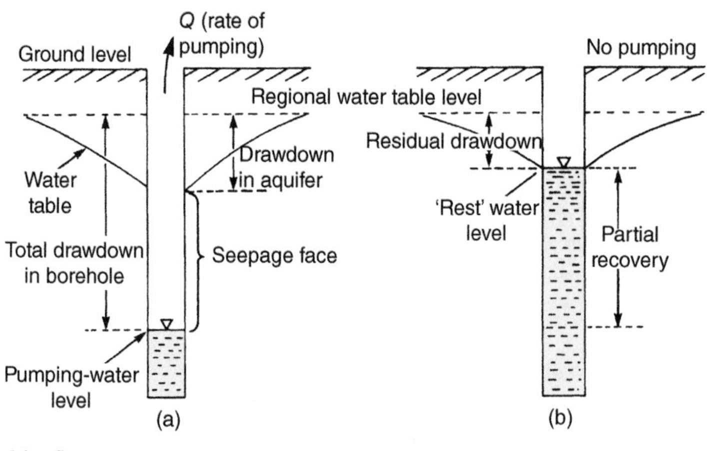

Figure 4.16 *Components of drawdown and recovery in an abstraction borehole.*

pumping level is achieved. The total drawdown below the regional water table is the sum of the drawdown in the aquifer, and the height of a seepage face in the well depends on the permeability of the well face and falls in the range 0.1–50 m. Once the pumping is stopped, the seepage face quickly disappears, but recovery within the aquifer takes considerably longer (Figure 4.16b).

As a result, water levels in the borehole are likely to be affected by residual drawdown for several days, weeks or even longer after pumping has ended. This means that it is important to know the pumping rate and how long the borehole was pumped before each measurement was made to try to make sense out of a pumping-water level record. It is sometimes possible to estimate general trends in water level fluctuations for long-term records. Plot separate graphs of rest and pumping levels, as in the example shown in Figure 4.17, and use the record of volumes pumped to see whether any trends have been caused by changes in abstraction. The pumping water level record varies, reflecting changes in the total quantities of water pumped from the borehole each year. In Figure 4.17, the higher water levels in 1974 and 1976 were caused by a reduction in the hourly pumping rate. The rest-water levels are fairly constant at some 30 m below datum. This indicates that the pumping from the borehole is not causing a general decline in the regional water level. It does not provide a record of the regional level, however, as the rest level measurements only represent a partial recovery, as shown in Figure 4.16b. The few

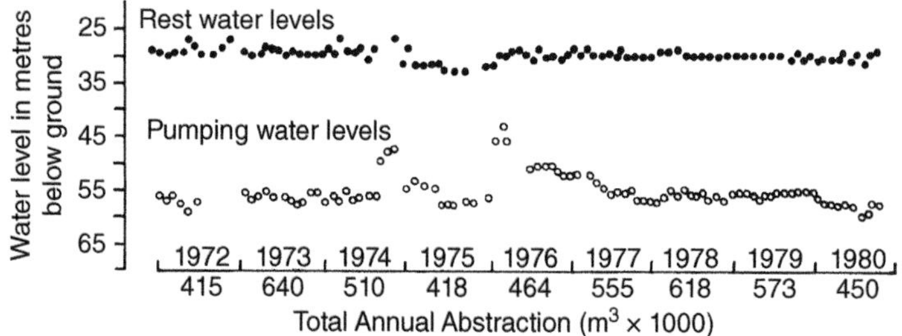

Figure 4.17 *This example shows a nine-year record of a factory borehole, where rest and pumping-water levels were recorded at monthly intervals. (Data by courtesy of the Environment Agency.)*

rest-water levels that are higher than the others were taken after rest periods of a few weeks, instead of the usual two-day weekend shutdown. This suggests that the regional level is several metres higher than the rest-water level record from this borehole. In general, the record shown in this example provides useful information about both the behaviour of the abstraction borehole and the impact of the abstraction on the local groundwater levels. This data set is not sufficient to provide much quantitative information, however, despite being more comprehensive than the majority of such records.

Try to use a standard recovery period of at least several hours and ideally overnight before taking the rest water reading in an abstraction borehole. You may be able to take advantage of periodic shutdowns in pumping, such as each weekend or during the annual holiday period. If all else fails, use the storage capacity in supply tanks to organise a few hours without pumping.

4.8 Groundwater-Level Monitoring Networks

A large number of measurement points are needed to define groundwater-level contours but with measurements taken perhaps only twice a year, in late spring and early winter when groundwater levels are at their maximum and minimum values, respectively. In contrast, monthly or even weekly readings are needed at fewer locations to define the changes in groundwater levels in response to seasonal recharge and abstractions, with only one borehole every 20–25 km^2 being enough for regional studies. A much greater density of observation points will be needed when you are examining a small area. For example, the investigation described in Case History 2 had a density of three piezometers per hectare.

Case History 2 – Groundwater-Level Interpretation

A risk assessment was undertaken to identify potential impacts on the water environment from a proposed new cemetery. The site is underlain by glacial sand that overlies Carboniferous mudstones. A major aquifer, the Sherwood Sandstone, has a faulted boundary with the mudstones to the northwest of the site and the sands extend across this fault to overlie the Sherwood Sandstone. The field is almost level at an elevation of about 113.5 m above Ordnance Datum (mOD) and falls on a watershed between two minor watercourses. A spring about 100 m to the south of the site forms the head of a minor stream and is the closest surface water feature. The preliminary conceptual model based on the topography and drainage supposed that the groundwater flow from the site would be to the south in the direction of this spring.

A series of six standpipe piezometers were installed round the periphery of the site using dynamic probing methods (see Section 9.1). Each is made from 19-mm diameter galvanised-steel piezometer pipe, with a 500 mm-long perforated tip driven to the maximum possible depth within the limitations of the installation method (approximately 5 m).

The top of each piezometer was levelled to a common datum using a surveyor's level (see Section 4.3.1), and the groundwater-level data were used to construct a water table contour map, shown in Box Figure 2.1, using the method described in Section 4.11. Before the piezometer data were available it had been assumed that the groundwater flow would be towards the spring in the south. Data for March 2004 were chosen as the highest levels recorded during the year. The water table contours show that the groundwater flows towards the northwest and not towards the spring. Records from a pumping station in the sandstone aquifer some 2 km from the site showed that groundwater levels have fallen by about 30 m at the pumping borehole since it was constructed in 1935. This implies that groundwater levels have fallen across the sandstone aquifer, thereby encouraging groundwater in the sands to drain into the sandstone aquifer, changing its flow direction away from the spring.

The water table contour map shows a steep drop in groundwater levels of about 0.5 m near the eastern side of the site. A deep sewer trench crosses the site in this location and the differences in levels indicate that the backfill material in the trench has a relatively low permeability that induces a steep water table gradient to drive flow across it.

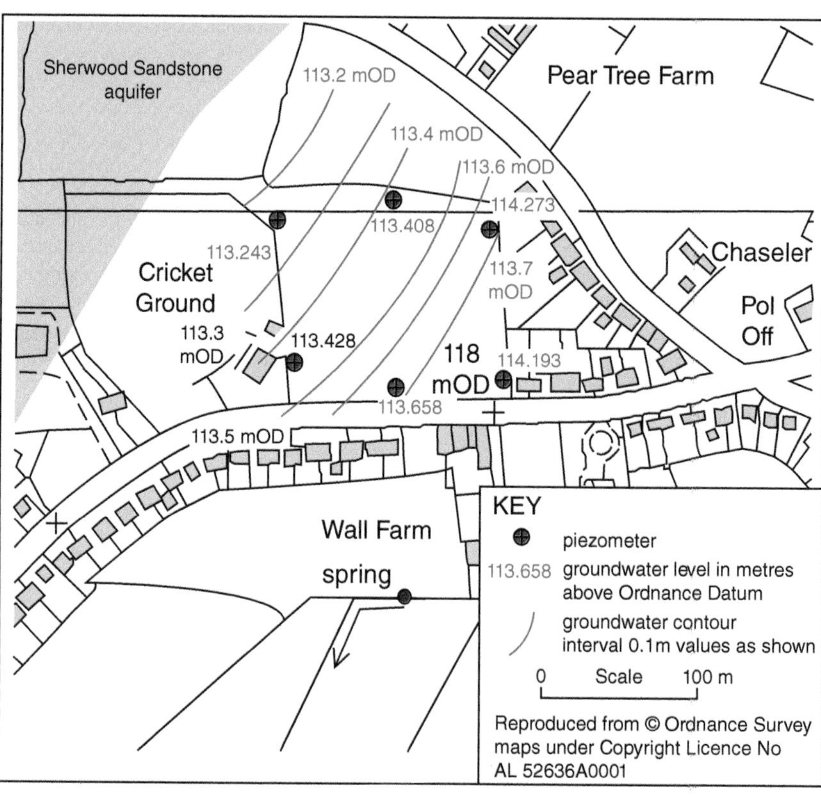

Box Figure 2.1 *The proposed cemetery is the field to the east of the Cricket Ground. The location of six piezometers is shown, as are groundwater contours at 0.1-m intervals. Contours show a flow direction to the northwest towards the Sherwood Sandstone aquifer where groundwater levels have been lowered by abstraction. (Data by courtesy of Newcastle-under-Lyme Borough Council.)*

The ideal frequency to take readings depends upon how quickly the aquifer responds to recharge, but it will normally take a year or two to find this out. As a rule of thumb, for general monitoring purposes take monthly readings and install water level recorders or data loggers on about 10% of the boreholes. More frequent readings are needed to monitor the impact of pumping as part of a test or that caused by operational pumping from a group of wells. Practical considerations usually justify the installation of data loggers. Alternatively, fit in your measurement programme with other work in the same area. Keep a dipper in your car and measure water levels when you are passing any of your observation boreholes, thereby obtaining valuable extra data for a minimum of effort.

4.9 Groundwater-Level Fluctuations

Groundwater-level data are used to define seasonal fluctuations caused by recharge, or the long-term impact of pumping. Other factors can cause water levels to change and must be taken into account in your interpretation of the data. The theory behind these other influences is described in all standard groundwater textbooks, such as those in the reference list. Groundwater recharge is discussed in some detail in section 3.6 Recharge through the Vadose Zone, which shows that a direct relationship between rainfall and rises in the water table rarely generally do not happen.

4.9.1 Changes in groundwater storage

The groundwater hydrograph for a borehole at Chilgrove House, West Sussex, in the unconfined Chalk aquifer of the South Downs is shown in Figure 4.18 and illustrates the seasonal fluctuations caused by rainfall recharge. Groundwater levels have been recorded here since 1836, making it the longest continuous record in the UK. The well is located in the Seaford Chalk Formation about 10 km north of Chichester. It consists of a shaft some 0.9 m in diameter to 43.74 m, with a 147-mm borehole, sunk to 62.03 m and cased from the surface to the bored section. Originally 41.15 m deep, it was cleaned out and deepened to 43.74 m in 1855 and further deepened in 1934 to 62.03 m.

Hydrographs for the period from 2010 to 2016 generally show higher than average values in winter periods and lower values during summer months; compare them with the long-term values to identify significant events. The wet winters of 2009/2010, 2012/2013, and 2013/2014 caused a rise

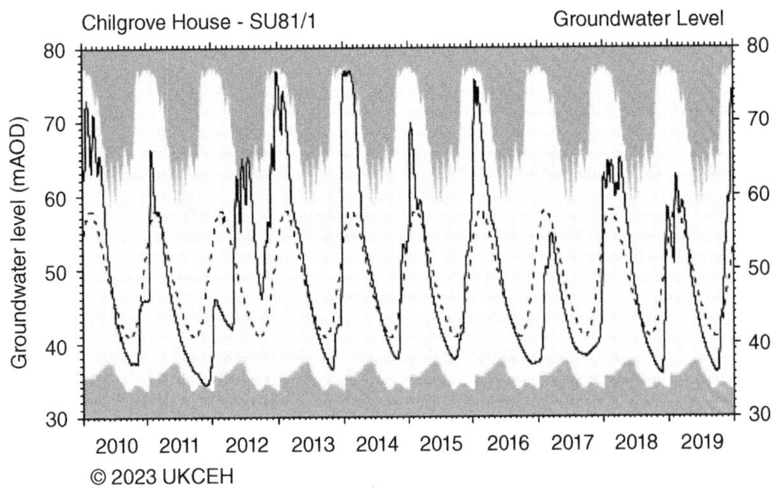

Figure 4.18 *Groundwater levels measured at Chilgrove House since 1836 is believed to be the longest continuous record of water levels in the UK, and possibly the world. The well is located on head deposits over Seaford Chalk Formation in a branch off the generally dry Chilgrove valley about 10 km north of Chichester. It consists of a shaft some 0.9 m in diameter to 43.74 m with a 147 mm diameter borehole, sunk to 62.03 m that is cased from the surface to the bored section. It was originally 41.15 m deep and was cleaned out and deepened to 43.74 m in 1855 after the drought of 1854. It was further deepened in March 1934 by a 147 mm diameter borehole to a depth of 62.03 m. The hydrograph is from 2010 to 2019. The solid line is the hydrograph and the dotted line is average values. The long-term maximum (blue) and minimum (pink) values are also shown. The hydrographs for the period generally show higher than average values in the winter periods and lower values during the summer months. The exceptions are the winter of 2011/2013 which has lower than average values and the end of 2012 that has higher values from recharge in the summer months. (Reproduced with permission from UK Centre for Ecology & Hydrology.)*

in groundwater levels to almost the highest recorded, whereas the dryer weather from the summer of 2011 until the early part of 2012 produced levels that were below average. Changes in water levels also reflect variations in abstraction rates and tend to follow much longer trends. Other changes in groundwater storage can occur in specific situations. For example, water levels in aquifers close to a river may fluctuate in response to changes in river level as water flows in and out of the riverbank deposits. The extent of these changes will depend upon the size of the river-level fluctuations and the permeability of the bank side and aquifer materials. Comparison of well hydrographs with the river hydrograph (see Section 5.7) will help to detect this relationship. A similar effect may be observed in boreholes right on the coast where the aquifer is in direct hydraulic contact with the ocean and where the weight of water moving on and off the aquifer has a pressure effect. Water level changes caused by water flowing in and out of the aquifer with the tide are only observed on the water's edge, although pressure effects are seen much further inland.

4.9.2 Barometric pressure

The groundwater surface in wells in confined aquifers represents a balance between the hydrostatic pressure in the aquifer and the weight of the atmosphere that fills the top of the borehole. Consequently, changes in atmospheric pressure produce a corresponding change in water level, with the water level falling as the barometric pressure increases and rising as the barometric pressure drops. A large number of boreholes exhibit this behaviour, even some that penetrate water table aquifers! In water table aquifers changes in atmospheric pressure are transmitted equally to the water table in both the aquifer and a well; hence no pressure differences should occur, in theory at least! Where fluctuations are seen there are several possible causes. For example, air trapped in pores below the water table will respond to changes in atmospheric pressure, causing small fluctuations in the water level, or the aquifer is confined by clay that may have elastic properties. Deep boreholes in water table aquifers with a marked anisotropy frequently show barometric responses.

Barometric changes need to be removed from a groundwater-level record so that other influences can be seen. To do this, it is first necessary to calculate the *barometric efficiency* of the aquifer. This must be done for each borehole, as local geological conditions can cause it to vary from place to place. Barometric efficiency is simply the change in water pressure divided by the change in barometric pressure in the same units, and is expressed as either a percentage or a decimal.

Figure 4.19 shows how to calculate barometric efficiency by plotting the two records so that they can be compared visually. Note that the vertical scale of the lower graph has been reversed to show the same pattern as the water level changes. The atmospheric pressure is also shown in centimetres of water rather than millibars to make the comparison easier. In this example, the change in water level and its corresponding change in atmospheric pressure were estimated from the graph and the water level change calculated as 23% of the barometric change. Experience has shown that the majority of barometric efficiency values fall in the range of 20–80%. The value of barometric efficiency is usually different for rising and falling pressure conditions.

An alternative method is to plot a graph of barometric pressure on the x-axis and water levels on the y-axis, as shown in Figure 4.20, which uses data from Figure 4.19. The slope of the straight line is the barometric efficiency expressed as a decimal or as a percentage. The generally poor fit of this plot is attributed to the barometric measurements being from a station some 18 km from the borehole. This distance will cause a time lag between the two records that will vary depending upon the prevailing weather conditions. The method also mixes rising and falling conditions and so produces an average value.

It is also possible to estimate the short-term barometric efficiency using the changes in barometric pressure and water level over a constant time interval of an hour or perhaps a day. Bear in mind that the barometric effects depend on several factors, some of which may change. It is not certain which

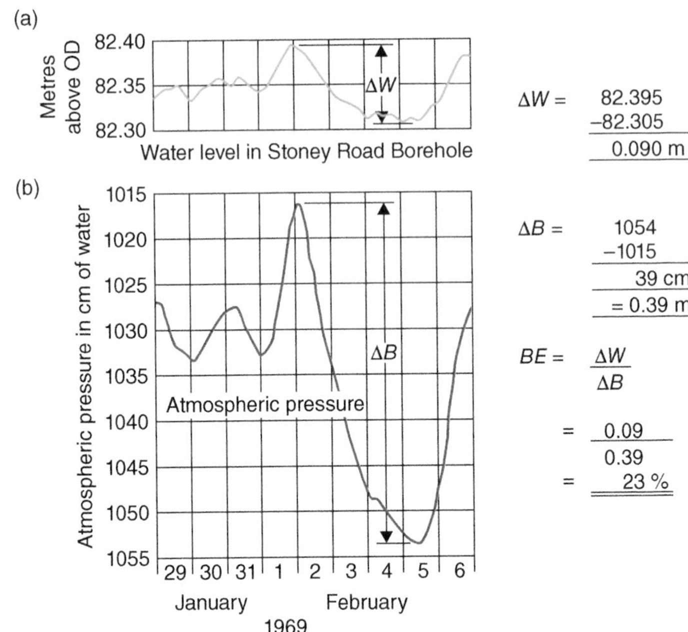

Figure 4.19 Groundwater levels (a) fluctuate in response to changes in atmospheric pressure (b). In the lower graph the atmospheric pressure is shown in centimetres of water and the vertical scale has been reversed to make the comparison easier.

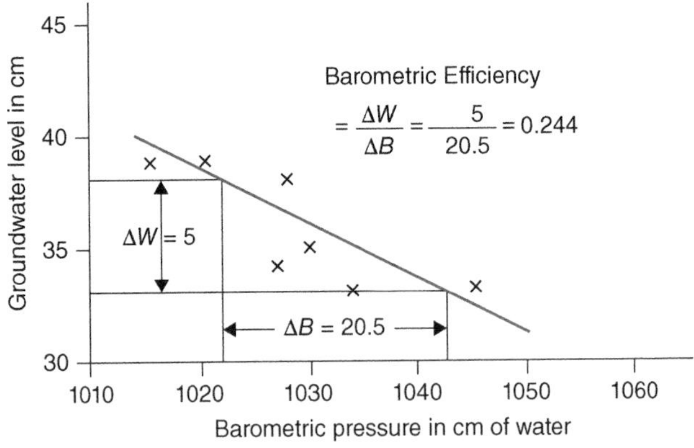

Figure 4.20 Barometric efficiency can be calculated by plotting water levels against atmospheric pressure (expressed as a column of water). The barometric efficiency is the slope of the straight line expressed as either a decimal or a percentage. (Note: based on data used in Figure 4.19.)

one should be used in correcting groundwater-level records, but in a semi-confined aquifer it is probably best to use the short-term barometric efficiency, as the long-term values may be too low. The slopes of these graphs will be affected by the recharge signal, which will affect the gradient of the plot if less than one year of data are plotted, which may often be the case. For records of a few weeks to a few months it is a good idea to de-trend the data first by calculating the linear regression line for the whole data set and then subtract it before the data are plotted. This will remove most of the effects of the recharge signal and should improve the estimate of the barometric efficiency. If you need to go into barometric efficiency in more detail look at the method developed by Clark (1967) which is designed to correct for influences like recharge and pumping effects. In addition, Toll and Rasmussen (2007) have developed a computer program that allows you to correct for barometric effects and those of Earth tides.

Once you have calculated the barometric efficiency you can correct your groundwater-level data by following the steps given below.

1. Determine the atmospheric pressure at the time each groundwater level was measured.
2. Convert these values to units compatible with your level record (e.g. centimetres of water, where 1000 mbar = 1019.7 cm of water at 4°C).
3. Calculate the difference in atmospheric pressure between each reading and the previous one and multiply these values by the barometric efficiency. This will give the size of the water level movement caused by the barometric change. Use the appropriate barometric efficiency value for falling and rising stages.
4. If the atmospheric pressure has increased from when one water level reading was made to the next, *subtract* the amount calculated in step 3 from the measured drawdown, where the water level reading is measured as a *depth* below the ground level. This is because increasing atmospheric pressure will have depressed the groundwater level. The reverse applies when the barometric pressure has fallen. Here, the correction value should be *added* to field-water level readings. Where groundwater levels are corrected to read *above* a datum (e.g. sea level), reverse the procedure, adding the correction value when atmospheric pressures increase, and vice versa.

4.9.3 Changes caused by loading

Confined aquifers have an elastic property and move in response to loading, causing changes in the hydrostatic pressure. In coastal areas, tidal fluctuations cause large changes in the mass of seawater over aquifers that extend beneath the ocean, with a corresponding change in pressure in the aquifer. These tidal fluctuations are direct; that is to say, as the sea level increases so does the hydrostatic pressure and the water levels in the wells rise. These effects can be recognised by their semi-diurnal nature and by comparison with tide tables, although a time lag is often observed.

Figure 4.21a shows a recorder chart from a tidally affected borehole. This borehole is in a confined aquifer and lies within 1.5 km of the coast. The regular fluctuations that occur at 12-hourly intervals are caused by the tide. The changing load of seawater on the aquifer causes pressure changes within the aquifer, which in turn cause a change in level. In this example, high tide occurred approximately three hours before the peak in water level.

The borehole in Figure 4.21b is in a confined aquifer and is located at Coventry railway station. The very noisy trace is caused by changes in loading on the aquifer as trains pass by. The overall changes in water level are caused by fluctuations in barometric pressure, data from this borehole also used in Figure 4.19.

The borehole in Figure 4.21c is in a confined aquifer and lies within 2.5 km of the coast in northwest England. The record of water levels shows the effect of both tidal movement and barometric fluctuations. The main feature on this trace, however, is a fluctuation in water level of

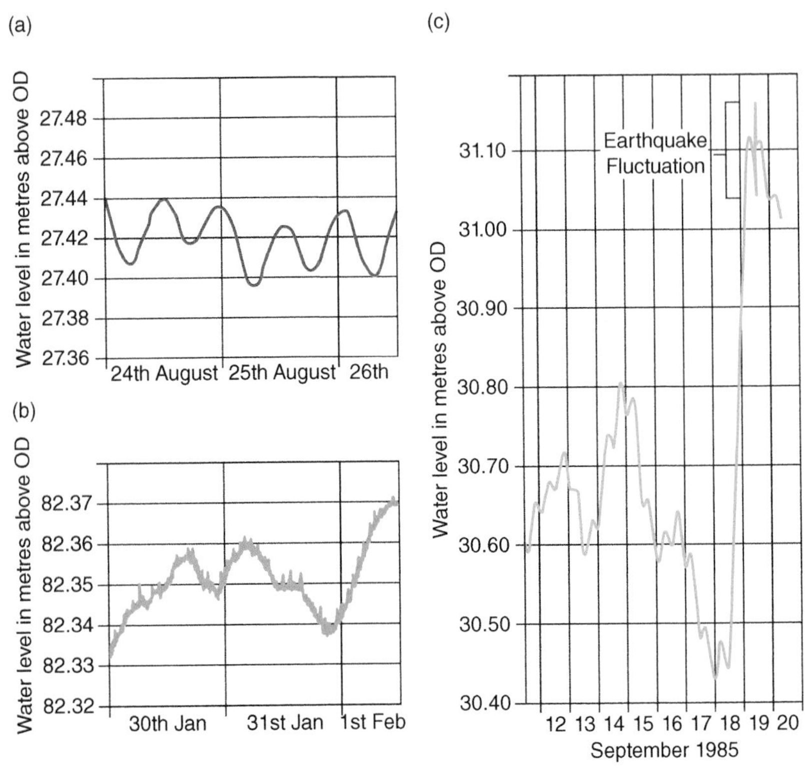

Figure 4.21 *The three graphs are copies of recorder charts on observation boreholes and show various external influences on groundwater levels. In (a) the borehole is about 1.5 km from the coast; in (b) the borehole is located at a railway station; and in (c) the water levels are affected by an earthquake on the other side of the Earth. (Data by courtesy of the Environment Agency.)*

some 12 cm caused by shock waves from the Mexican earthquake on 19 September 1985 with a magnitude of 7.8 on the Richter scale.

Detailed measurements of tidal fluctuations near the coast can produce data that are difficult to interpret because of the way water flows in and out of the aquifer.

The diurnal fluctuations shown in the diagram are entirely caused by changes in the loading on the aquifer by seawater. However, water table aquifers on the coast are often affected by changes caused by the flow of water, as shown in Figure 4.22. In an investigation illustrated here, groundwater levels were monitored in a layered aquifer next to a tidal estuary using piezometers at two different depths. Water levels were recorded at frequent intervals over a tidal cycle that showed that the lower piezometer (B) was affected with a small time-lag from the tide, whereas the upper piezometer (A) did not change at all. These field measurements were interpreted as the lower piezometer reflecting a loading response. The upper piezometer, at a higher level in the aquifer, reflects only water table responses. The distance between this piezometer and the estuary was too far (10 m) for the water flowing into the aquifer to alter the water level in the measuring device. It was assumed, however, that the groundwater stored in the banks fluctuated over a tidal cycle, as indicated in the diagram. The time of each high tide advances almost an hour a day, and so a simple comparison of

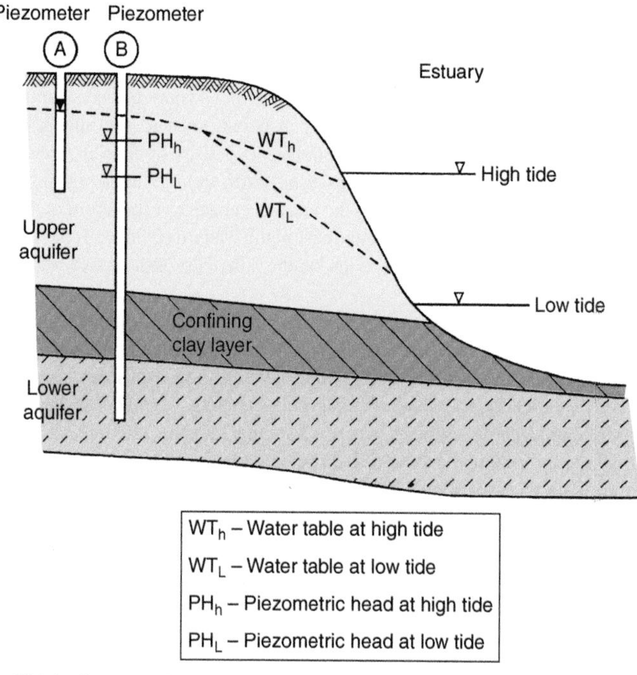

Figure 4.22 *Tidal effects can be assessed by comparing water level readings from piezometers close to the coast.*

the groundwater-level fluctuations with the tide over a period of a month will decide whether any regular daily fluctuations are tidal or have another cause, such as local abstractions.

Point loading on an aquifer also results in groundwater-level movement, with perhaps the most common example being the movement of railway trains. As the train approaches a borehole in a confined aquifer the well water level rises in response to the increasing load. Once the train has passed, the water level falls below the undisturbed level, before returning to rest again. These changes occur over a very short period, about the same time that the train takes to pass by. As a consequence, the passage of trains usually shows up as 'noise' on recorder-chart traces (see Figure 4.21b), and may be missed by water levels recorded on a data logger system unless it is set for very rapid readings. The amount of change in a groundwater level is rarely more than a few millimetres, and so this phenomenon is rarely more than a curiosity, once the cause of the noisy trace has been identified. A case history showing the effect of trains was described by Jacob (1939).

4.9.4 Miscellaneous

There are a variety of other causes for groundwater-level fluctuations. Earth tides are movements of the Earth's crust in response to the gravitational pull of the Moon and, to a lesser extent, the Sun. These movements have been shown to produce semi-diurnal water level fluctuations in some confined aquifers, and can be at any distance from the coast. Groundwater levels may fluctuate by several centimetres in confined aquifers; for example, an amplitude of up to 4 cm is seen in the confined Chalk aquifer of East Yorkshire, although the general range for all aquifers is 2–100 mm. Fluctuations caused by Earth tides may be easily detected using pressure transducers that are recording on an hourly or shorter interval. Such fluctuations can be puzzling if you are not familiar with the potential effects of Earth tides. The effect is observed most clearly in confined and semi-confined aquifers, especially those that are made of unconsolidated materials.

Earthquakes can cause changes in groundwater levels in a variety of ways. In the area directly affected by the earthquake, spectacular changes can occur, such as the appearance and disappearance of springs, fluctuations in spring discharges and rises or falls in groundwater levels. A permanent rise in the water table can be caused by Earth tremors inducing further compaction of non-indurated sediments, thereby reducing the storage capacity of the aquifers. Most frequently, however, earthquake shocks cause small fluctuations in the water level in wells penetrating confined aquifers. Such responses can be seen in boreholes on the other side of the world from the earthquake epicentre, as shown in Figure 4.21c.

The trace of a recorder chart can fluctuate for many reasons other than changes in water level. Wind gusting over the open top of a borehole causes a sudden drop in air pressure inside the casing, which then produces an immediate rise in the water level. Strong winds can give rise to a 'noisy' chart more directly by rocking poorly secured recorder cabinets. Animals can also affect the water level record. For example, cattle or horses may use a cabinet housing a float recorder as a scratching post, thereby rocking both cabinet and recorder, causing an unexplained wiggle of the pen trace. If this is done by cows on the way to be milked, a regular semi-diurnal feature may result. Any small animal that falls into a well that contains a float is likely to climb onto it and all the animal's movements will be transmitted to the recorder. Records of water levels made with data logger systems are far less susceptible to this type of confusing problem.

Groundwater levels may also be affected by a number of other factors such as leaks from water supply pipes and sewers. Evapotranspiration from deep-rooted plants or in areas with a shallow water table may lower groundwater levels, particularly during the summer. The natural variations in evaporation rates throughout a 24-hour period can result in a diurnal fluctuation of the groundwater levels. Crop irrigation, on the other hand, is likely to artificially increase recharge and has caused shallow water tables to rise even to the point of water-logging soils and surface flooding.

4.10 Managing Groundwater-Level Data

A groundwater-level data set should be recorded to a common datum system. In the UK, for example, levels are usually *reduced* (i.e. converted) to Ordnance Datum (approximately sea level), which is the system of levels used on Ordnance Survey maps. Measure the water levels in each borehole from the same fixed datum point such as the top of the casing. The elevation of these datum points should be measured using a surveyor's level (see Section 4.3.1) to the same level. The groundwater-level readings are reduced to a common level by *subtracting* the depth reading from the datum point value for that borehole. If the borehole is artesian it may be necessary to *add* the two levels. Your common sense will easily enable you to decide whether to add or subtract.

It is usual for the groundwater-level records to use reduced values. The raw data on field sheets or in electronic files are usually saved either for the duration of the project or, in the case of routine measurements, for a fixed period of time such as five years. This precaution allows the original record to be examined if there are any queries when the data are being used. The records can be kept as a series of numerical values for each site or in the form of a hydrograph. It is common for both to be used if a paper system is employed. Where data are stored in a computer database, hydrographs can be printed when they are needed. There are some commercially available databases that are designed for groundwater-level data, and these are particularly suited to data collected using a data logger system. The program can be easily set up so that the field data are reduced to a common datum and the data printed out as hydrographs or used to produce a map. Alternatively, you can use a standard spreadsheet package such as Microsoft Excel and create your own.

4.10.1 Data validation

Field data should be systematically checked to verify that they are valid readings. Mistakes occur in taking field readings and it is important to make sure that the data are accurate soon after they were taken so that you can remember any factors that may affect their accuracy. A simple method is to plot hydrographs for all the boreholes in an area and compare them to see if they follow the same pattern. If one well shows the water level to have gone down when the others have risen, either the record is wrong or the water level is affected by local factors such as pumping. When the level has risen when the others have fallen it could be a field error, instrument failure or a local increase in recharge caused by a burst water main, for example. Revisit the site to check out any irregularities. Groundwater levels from the same area but in different aquifers can be expected to follow the same general pattern, although the amplitude is likely to be different.

4.11 Constructing Groundwater Contour Maps and Flow Nets

Groundwater-level data are used to determine the direction of groundwater flow by constructing groundwater contour maps and flow nets. A minimum of three observation points is needed to identify the flow direction. The procedure is first to relate the groundwater field levels to a common datum – map datum is usually best – and then accurately plot their position on a scale plan, as shown in Figure 4.23. Next, draw a line between each of the observation points, and divide each line into a number of short, equal lengths in proportion to the difference in elevation at each end of the line. In the example shown in Figure 4.23, each division on the lines A–B and B–C is 0.2 m, while on line A–C each division is 0.1 m. Remember that the length of the divisions on each line is not related to the scale of the plan but to the difference in level between the points joined by the line. The next step is to join points of equal height on each of the lines to form contour lines.

This simple procedure can be applied to a much larger number of water level values to construct a groundwater-level contour map, such as the one in Figure 4.24. First, locate the position of each

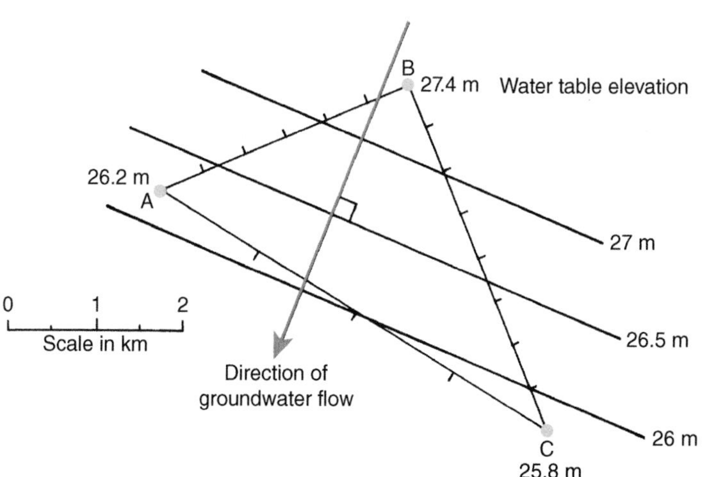

Figure 4.23 An illustration of how groundwater contours and flow directions can be estimated from groundwater levels measured in three observation boreholes.

observation point on a base-map of suitable scale, and write the water level against each position. Study these water level values to decide which contour lines would cross the centre of the map. Select one or two key contours to draw in first. The 22-m contour in Figure 4.24a was drawn by interpolating its position between each pair of field values following the procedure illustrated by Figure 4.23. Once this was completed, the 26-m contour was drawn in the same way. The remaining contours were drawn by interpolating between the field values using the two key contours as a guide. Select a contour interval that is appropriate to the overall variation in water levels in the study area.

Once the contour map is complete you can construct flow lines that are at right angles to the contour lines. Choose a contour line that crosses the middle of the area and divide it into segments of equal lengths. In the example shown in Figure 4.24b, the 30 m contour was divided into 500-m intervals as the starting point for constructing the flow lines. Draw a line from each point in a down-slope direction at right angles to the contour. When the next contour is intercepted the flow line is continued, but at right angles to this new contour line, until the next contour is reached and the process repeated. Always construct flow lines in a down-slope direction.

Bear in mind what you have already learned about the aquifer while you are constructing a groundwater-level contour map and flow net and take account of the geological structure and variations in aquifer hydraulic properties. This is particularly important where groundwater flow is through fractures such as karstic limestone. Fracture flow systems are very complex and actual flow directions are not easy to deduce from the groundwater-level contours. Groundwater tracer experiments are used to understand flow in such systems (see Section 9.3).

A contour and flow line map represents groundwater movement in a plan form only, and therefore is only part of the picture. When viewed in cross-section, groundwater flow paths curve

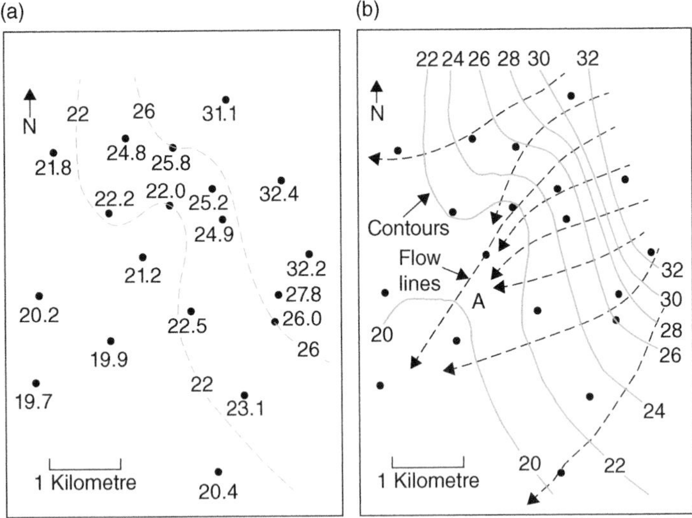

Figure 4.24 Construction of a groundwater contour map and flow net. (a) Values of groundwater levels are located on a plan, and key contours are plotted using techniques shown in Figure 4.23. In this example, contours at 22 m and 26 m were used. (b) Remaining contours are interpolated using these two contours as a guide. Flow-lines were sketched in, perpendicular to contour lines, starting on the 30-m contour at a spacing of 500 m.

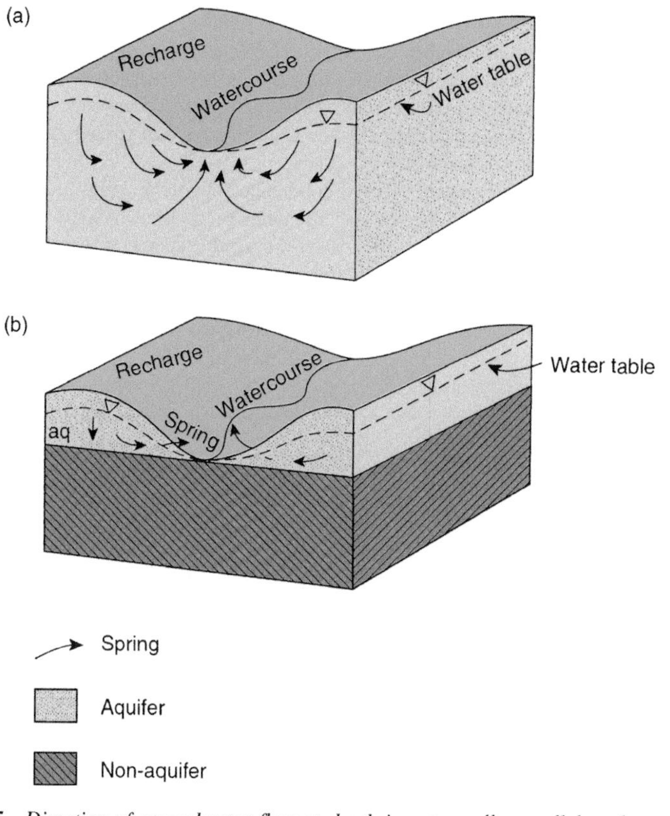

Figure 4.25 *Direction of groundwater flow at depth is not usually parallel to the water table; instead, water moves in a curved path, converging towards a point of discharge. Here two examples are shown, one in a deep aquifer (a) and the other in a shallow one (b).*

towards a discharge point such as a spring line, stream or even a pumping well (see Section 4.7). Figure 4.25 shows two examples of the 3D groundwater flow at depth in an aquifer. In Figure 4.25a, the ground is uniformly permeable and groundwater discharges into streams along the valleys; it may approach the stream from the sides or from below. In recharge areas, this downward flow would be reflected in the relative levels in a nest of piezometers, where the water level would be highest in the upper piezometer and lowest in the deepest one. The 3D distribution of heads in a recharge area is reflected in the water levels measured in the piezometers set at different depths shown in Figures 4.13 and 4.14. In discharge zones the reverse is true, and this accounts for flowing artesian boreholes being common in river valleys. In Figure 4.25b, the hill is capped by a permeable rock that is underlain by an impermeable stratum. Groundwater is diverted laterally by the impermeable material, and springs result at the ground surface along the contact between the permeable and impermeable strata. Try to picture in your mind the 3D groundwater flow paths in the aquifers you are studying as part of the process of developing your conceptual model.

4.12 Interpretation of Contour Maps and Flow Nets

A groundwater contour map can represent either a water table (phreatic surface) or a piezometric surface, and this should be sorted out from geological and well construction information. It is possible that one part of an aquifer is confined while the remainder has a water table (see Figure 1.2b). Indeed, such complexities are common but must be understood before groundwater contour maps and flow nets can be interpreted accurately.

The spacing of groundwater contours gives a good indication of variations in the aquifer transmissivity values. Where contours are close together it indicates low transmissivity values, because a steep hydraulic gradient is needed to drive the water though the aquifer. Where groundwater contours are more widely spaced, the converse is true, and the transmissivity values will be higher. Transmissivity reflects both the permeability and aquifer thickness, so these differences in the gradient of the groundwater surface may reflect changes in either aspect or both.

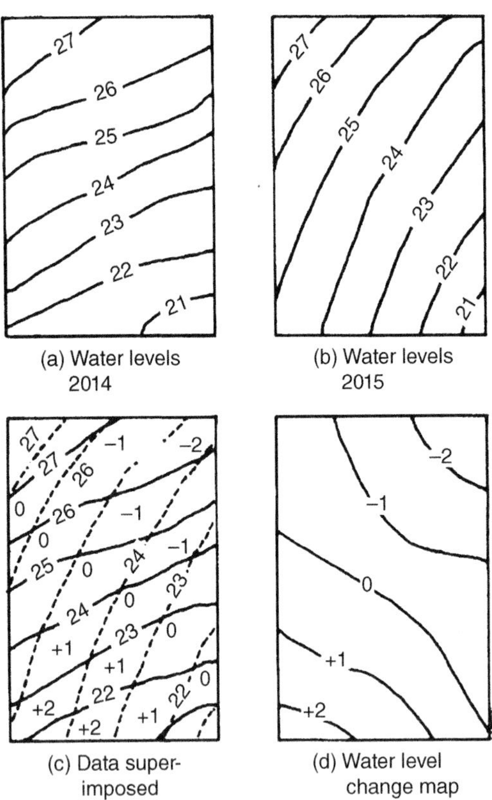

Figure 4.26 *Changes in groundwater levels can easily be studied if you construct a water level change map.*

Groundwater flow lines indicate not only the overall direction of flow but also where flow is concentrating, again reflecting variations in transmissivity. In the example in Figure 4.24b, the groundwater contour spacing suggests that the aquifer is more permeable in its southwest part than in the northeast section or possibly the aquifer is thinning to the northeast. The flow lines show that there is a concentration of flow in the area marked 'A', and this is probably an area of discharge to surface streams. If this map were to be used to select favourable locations for a new well, area 'A' would have much to commend it as it is both in an area of high permeability values and where flow lines converge.

Water level change maps (Figure 4.26) are used to calculate changes in volume of water stored in an aquifer and are part of the water balance exercise (see Chapter 8). In the example in Figure 4.26, groundwater-level contour maps (a) and (b), covering the same area in successive years, are superimposed in (c). The differences in values are noted at points where the contours cross, and new contours are drawn to show the amount of change (d). Water level change maps are useful to measure the local effects of recharge and discharge. In this example, groundwater levels rose in the southwest part of the aquifer and fell in the northeast. These differences are likely to be caused by factors such as variations in abstraction or recharge. These maps may also be useful when assessing the local effects of recharge and abstraction. Case History 2 shows how groundwater-level contours can be used to deduce a significant amount of information on groundwater flow that is not evident from geological and surface water information.

4.13 Using Other Groundwater Information

Typically, there are fewer groundwater-level measurement points than the hydrogeologist would like. The construction of new boreholes to provide this extra information is very costly and so most experienced hydrogeologists have learned to use indirect information on groundwater levels. The position and elevation of springs can be used *provided* that there is evidence to show that these springs discharge from the aquifer you are studying. Groundwater chemistry is useful here using the method discussed in Chapter 7 and in Case History 2.

Streams and rivers can also be used to help construct a groundwater contour map. Field evidence is again needed, and this may include outcrops of the aquifer in the streambed, with a significant increase in flow as the stream flows across the aquifer. Other commonly used information includes topographic details taken from published maps. If the study area is on the coast, it is usual to assume that the coast approximates to the zero groundwater contour. Similar assumptions can be made in respect of any large lakes in more inland locations. It is important not to forget, however, that the water level in both water bodies fluctuates with the tide or the input of surface water, and that may need to be taken into account in your interpretation.

Figure 4.27 shows the groundwater contour map drawn up as part of a groundwater resource appraisal of the St Bees Sandstone aquifer in West Cumbria, England. Groundwater levels were only available at 11 sites that were unevenly distributed over the 30 km^2 of aquifer. This information was supplemented by a number of springs which all lie in the northern part of the aquifer. Similarity between the chemistry of the water from these springs with that of groundwater samples taken from the observation boreholes demonstrated that the springs drain from the main aquifer and not the overlying drift material.

Flow gauging (see Section 5.4) was carried out on the four streams that cross the aquifer to ascertain where the groundwater discharge points lie. This information was taken into account when the contour map was constructed, together with the shape of the local topography and an assumption that the coast approximates to the zero contour. Despite this additional information, it was not possible to draw contours with a more frequent interval than 10 m, and there was insufficient detail to allow the 10-m contour to be drawn over much of the aquifer area. Nevertheless, even with these

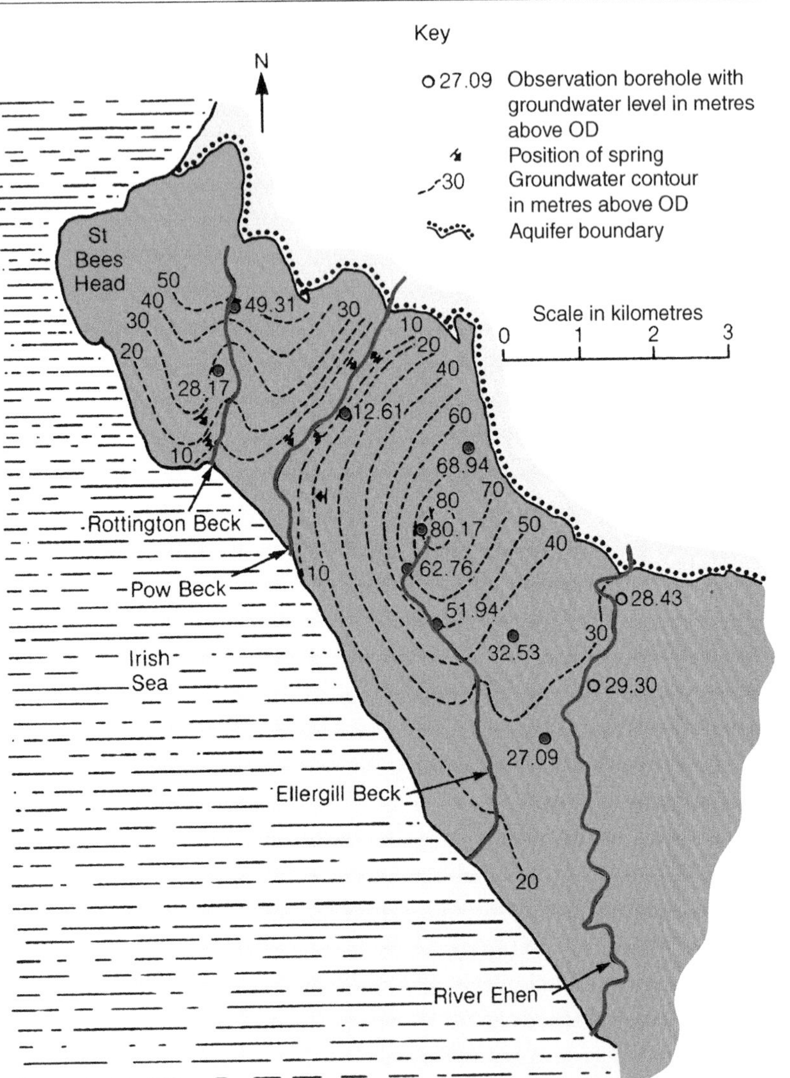

Figure 4.27 Groundwater contour map for part of the St Bees Sandstone (Permian) aquifer in West Cumbria, England (see Figure 2.2). (Data by courtesy of the Environment Agency.)

limitations, the contour map indicated that the northern part of the aquifer would be unlikely to be suitable for production boreholes, in contrast to the southern end. Note the contrast in the steepness of the groundwater 'surface' and the variation in the distribution of data points. These conclusions were supported by subsequent pumping tests.

5

RAINFALL, SPRINGS AND STREAMS

This chapter describes how to measure the other components of the hydrological cycle. Quite often you may use existing records of these parameters rather than making your own measurements. Use the information given in this chapter to check their accuracy and suitability for use in your project.

5.1 Precipitation

The majority of precipitation falls as rain in most parts of the world. Rainfall is measured by a network of rain gauges operated by national and local government agencies and it may be that one of these rain gauges is close enough for you to use these records. Where no existing gauges are available it will be necessary to set up your own. Sometimes several rain gauges may be needed to supplement sparse records or examine local variations. Differences over relatively short distances can be significant in areas of high relief where sites are at very different elevations. For example, there may be twice as much rainfall at the top of a hill as at its bottom, and this may be significant in studies of spring discharges.

A 'standard' Meteorological Office (UK) daily rain gauge (Figure 5.1) consists of a brass cylinder that incorporates a funnel leading into a collection bottle. The instrument is set into the ground to keep it stable and to keep the collected rain water cool to minimise evaporation.

If you cannot get hold of a standard rain gauge you can make one using a flat-bottomed glass bottle and a funnel or even a modified plastic bottle (Figure 5.2). Measurements are made at the same time each morning, and the daily total attributed to the *previous* day and recorded, as shown in Figure 5.3 below. The improvised rain gauge shown in Figure 5.2a uses a plastic funnel, glass drinks bottle and a large plant pot. The volume of rainwater collected in the bottle is measured using a measuring cylinder in millilitres and then converted into a depth using the method explained here. Remember that *millilitres* are equivalent to *cubic centimetres*, and that you want to record the rainfall as a depth in *millimetres*. Take great care. *If you get it wrong, you may end up being out by a factor of 10.* The water collected in a rain gauge is measured to an accuracy of 0.1 mm using a calibrated cylinder specially graduated in millimetres of rainfall and related to the rain gauge funnel diameter. Either pour the collected rainwater into the measuring cylinder on site or use a temporary holding jug to carry it to a place where the measurement can be carried out more easily, such as in a nearby building. If you are using a home-made rain gauge, measure the volume of rainwater to the nearest millilitre using a laboratory-type measuring cylinder or the sort used for home brewing.

Rainfall is expressed as a depth in millimetres and is calculated from the formula:

$$h = v / 10\pi r^2$$

where h is the depth of rainfall in *millimetres*, v the volume of rainwater in *millilitres* (which are the same as cubic centimetres) and r the radius of the funnel in *centimetres*. Record the rainfall in millimetres, rounded *down* to the nearest 0.1 mm.

Field Hydrogeology, Fifth Edition. Rick Brassington.

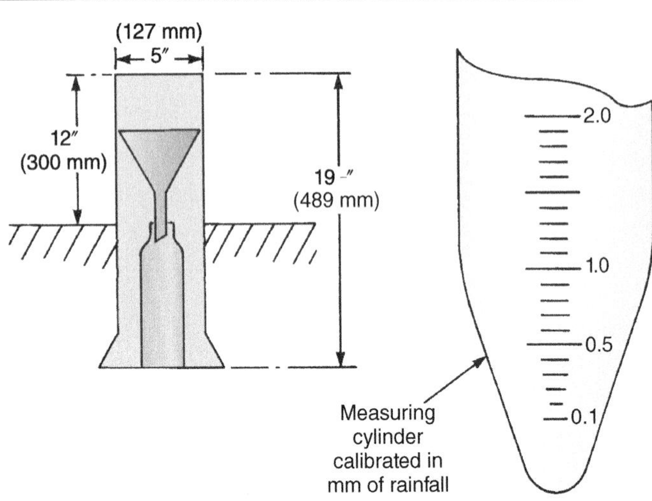

Figure 5.1 *A 'standard' Meteorological Office (UK) daily rain gauge consists of a brass cylinder that incorporates a funnel leading into a collection bottle. The volume of rainwater is measured each day using a standard measuring cylinder calibrated in millimetres depth of rainfall.*

A second improvised rain gauge (Figure 5.2b) has been made from a large plastic drinks bottle, cut in two to make both a funnel and the collection jar. Cut the plastic carefully to get a straight edge for the funnel. It is likely that you will have to hold the funnel in place with sticky tape and partly bury the base of the bottle to prevent it being blown over in the wind.

Rainfall amounts are always expressed as a depth of water in millimetres rather than as a volume. Records are kept as daily totals, with measurements being taken at the same time each day. In Britain, this is traditionally at 09.00 GMT, with the total being recorded as the rainfall for the *previous* day. It is standard practice to record the daily total to the nearest 0.1 mm *at* or *below* the bottom of the meniscus (water surface) in the measuring cylinder. If the volume of rainfall is less than 0.1 mm it is recorded as a *trace*.

Once you have measured the volume of rain collected in the gauge, shake out any remaining drops of water from the bottle. If it is raining when you make the measurement, empty the collecting bottle into the measuring cylinder or holding jug as quickly as possible, and then return it to the rain gauge. Any rain that you miss during that period is unlikely to be significant. Remember that you are recording the volume of rain that falls during a fixed period of 24 hours, so that the remainder of that shower will be logged with the rain that falls the next day.

If you miss reading the rain gauge for a day or two, try to maintain as accurate a record as possible. Record 'nil' rainfall if you know it is quite obvious that there has been no rain (or snow, hail, etc.). Enter '*tr*' for trace, if you know that there has been some rain, even though the bottle may be bone-dry when you take the reading. Record any collected rain as an accumulated total for the days since you took the last reading. Bracket the days together on your record sheet, but make a note of any showers. Figure 5.3 shows how these records are kept.

Positioning a rain gauge is very important. The location is usually a compromise between being out in the open, so that the gauge is not sheltered by trees or buildings, and exposed sites where strong winds could cause near-ground turbulence that might blow raindrops out of the mouth of the gauge. A rain gauge should be set at a distance from trees and buildings which is equivalent to a least

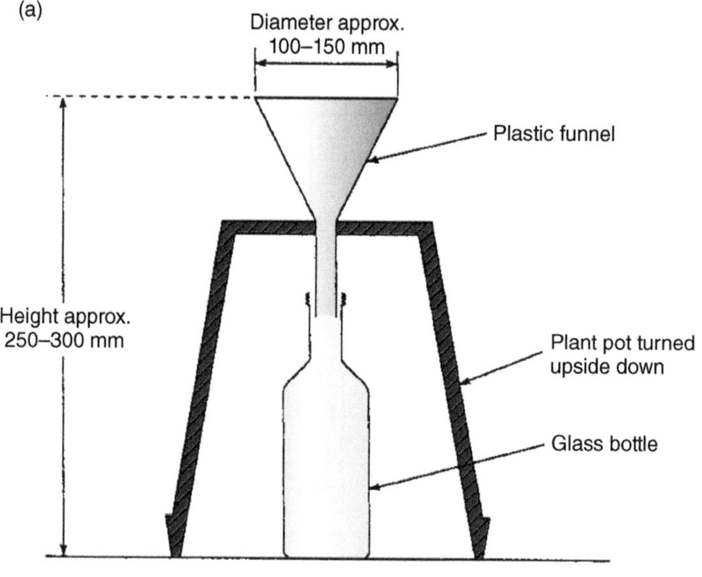

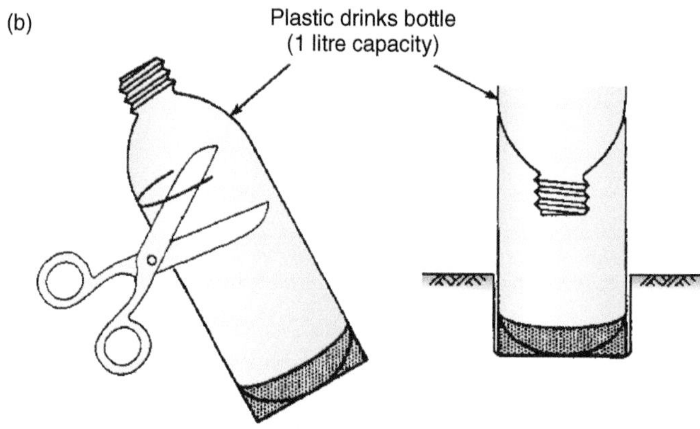

Figure 5.2 *Examples of improvised rain gauges that should be sufficiently accurate for most groundwater studies.*

twice their height, that is, the vertical angle between the gauge and the top of the trees or building should be no more than 30° (see Figure 5.4).

Snowfall is recorded as an equivalent depth of rain, that is, in millimetres of *water*. Collect a representative depth of snow in a cylinder and then carefully melt it and measure the volume of water produced. Avoid losing water from evaporation by standing the cylinder in a bowl of warm water in a cool room and allow the snow to melt more slowly. Snowfall measurements are difficult and subject to many more errors than rainfall measurements. When taking a sample, avoid both

Stn. name East Side Farm Month March Year 2006

Enter amount measured at 9h GMT against YESTERDAY'S date

Date	mm	*Enter time of measurement if not close to 9h GMT and notes on significant weather*	Date	mm	*Enter time of measurement if not close to 9h GMT and notes on significant weather*
1	2.0		16	1.1	
2	TRACE		17	5.7	HAIL
3	1.7	SNOW	18	0.6	
4	TRACE		19	1.0	
5	3.3		20	2.4	SNOW
6	7.1	RAIN / SNOW	21	–	
7	7.0	SNOW	22	5.4	SLEET IN AM.
8	–	THAW MIDDAY	23	4.3	SNOW BEFORE DAWN
9	–		24	10.6	HEAVY SHOWER LATE AM.
10	–		25		
11	–		26		SLEET IN AFTERNOON
12	TRACE		27	8.0	(Acc. Total)
13	TRACE		28	0.1	
14	0.5		29	0.6	
15	0.3		30	–	
			31	4.3	
			TOTAL	66.0	

OBSERVER H. Carson

Figure 5.3 *Make notes of significant precipitation events with your record of rainfall data. No readings were taken for the 25th and 26th; instead the rainfall for these days has been included with that for the 27th as an accumulated total.*

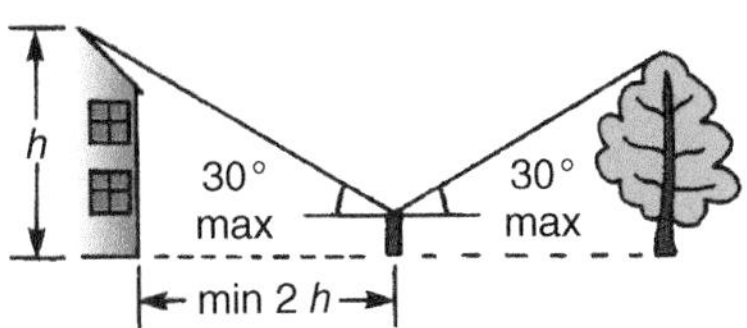

Figure 5.4 *Use the guidelines contained in this diagram to position your rain gauge so that measurements are not affected by turbulence caused by the proximity of buildings or trees.*

accumulations and 'thin patches' if the snow has been drifting, and do not collect snow left lying around from earlier falls. If snow forms a significant part of the total precipitation, you will have to carry out a snow survey. This is a systematic measurement of snow thickness and densities carried out along traverse lines. The information so gathered can be used to calculate the total volume of water represented by the lying snow which is then used to calculate the rainfall equivalent.

Calculating the total amount of rainfall over a catchment or recharge area, based on a number of rain gauges, was proposed by Alfred Thiessen, who worked for the US Weather Bureau in the early twentieth century. The method uses a series of polygons constructed round the pattern of rain gauges as shown in Figure 5.5 and is much more accurate than a simple arithmetic mean of all the rainfall measurements (Thiessen, 1911). To construct Thiessen polygons first connect the rain gauges by a series of lines. At the midpoint of each line draw a perpendicular line and extend them all to form polygons. Catchment rainfall is calculated by assuming that the average rainfall over each polygon is measured by the rain gauge at its centre. The total volume of rainwater falling on the catchment (or study area) is calculated by multiplying the area of the polygon that falls inside

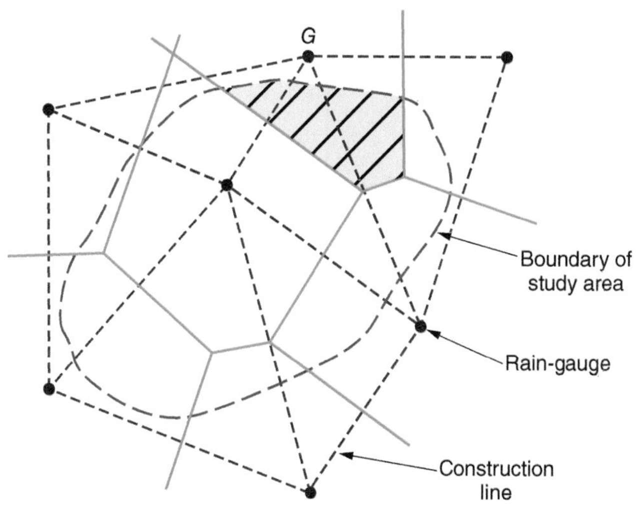

Figure 5.5 *Thiessen polygons are constructed to calculate the total amount of rain falling on a catchment.*

the catchment by the depth of rainfall at each rain gauge. For example, the volume of rain falling on the shaded portion of the catchment is taken to be the depth of rain measured at rain gauge 'G', multiplied by the area shaded. An easy way of measuring the area of each polygon is to trace it onto squared paper and count the squares. Then use the scale of the map to calculate the actual area as a proportion of the total area of the catchment. The total rainfall on each polygon is that value times the measured depth of rain. Of course, you could always use a computer program!

5.2 Evaporation

Evaporation measurements rarely form part of a hydrogeological field investigation where published figures are available, as these are usually adequate for most water balance purposes. Where data are not available, you will have to make your own measurements. It is most easy to measure evaporation as losses from open water. As these losses are the maximum that can occur, they will only give an idea of the potential losses from other surfaces such as soil or from plants.

The British Standard evaporation tank and the US Class-A evaporation pan are the types in general use (see Figure 5.6). The British Standard evaporation tank (Figure 5.6a) is 6 ft (1.83 m) square and 2 ft (0.61 m) deep, is made of galvanised iron, and is set into the ground with the rim 4 inches (100 mm) above ground level. The US Class-A evaporation pan (Figure 5.6b) is circular with a diameter of 47.5 inches (1.21 m) and is 10 inches (225 mm) deep. It is set 6 inches (150 mm) above the ground on a wooden platform so that air can circulate around it.

In both cases the tanks are filled with water, which is kept topped up on a daily basis. The water level in the British tank is not allowed to fall lower than 100 mm below the rim. Both types suffer from excessive losses during very hot conditions. Evaporation losses are calculated by carefully measuring the water level in the tank at the same time each day, and allowing for any water added to the tank to keep it topped up and also for any rain that may have fallen into it. Evaporation losses are expressed as a depth in millimetres of rainfall equivalents.

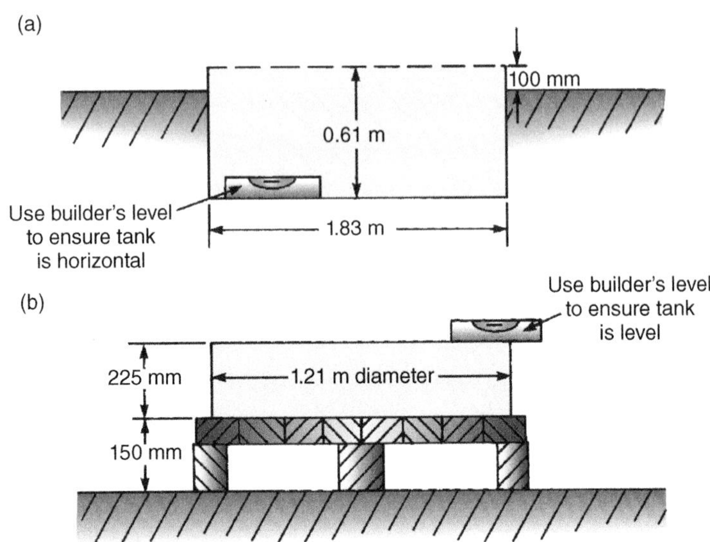

Figure 5.6 *Main features of a British Standard evaporation tank (a) and a US Class-A evaporation pan (b).*

August	Observations made at 0900 GMT		Daily records	
	Rainfall (mm)	Evaporation tank water level (mm)	Rainfall (mm)	Evaporation (mm)
1	1.2	23.6	–	1.5
2	–	22.1	10.2	0.7
3	10.2	31.6	–	1.8
4	–	29.8	8.1	2.3
5	8.1	35.6	–	1.6
6	0.4	34.0	0.4	2.0
7	–	32.4	*	*

Figure 5.7 *An example of daily readings of rainfall and evaporation losses. Note asterisks mean not yet calculated.*

Figure 5.7 shows daily readings of rainfall and evaporation-tank-water levels taken at 09.00 GMT. Rainfall is attributed to the previous day and the difference in tank-water levels is used to calculate the evaporation losses. For example, there was no rain on 1 August so evaporation is calculated as follows:

tank level on 1st 23.6
tank level on 2nd 22.1
therefore, fall due to evaporation = **1.5 mm**

When rain has fallen, this must be taken into account as follows:

tank level on 5th 35.6

tank level on 4th 29.8
therefore, rise in level = **5.8 mm**
rainfall on 4th 8.1
rise in level 5.8
therefore evaporation = **2.3 mm**

If you build your own evaporation tank make sure that it is not smaller than those shown in Figure 5.6, as small volumes of water warm up easily, which will increase evaporation rates. It is important to have a tank with vertical sides, so do not use your old bath tub! Straight sides mean that the water surface area remains constant and the calculation of water losses is kept simple. It must be set with the rim *absolutely horizontal* – so use a builder's level – because the measurements use the rim as a datum point. Make sure that birds and animals do not use it as a watering hole! Use wire netting to cover the tank and a fence to keep animals away.

5.3 Springs

The simplest way of measuring the flow of most springs and sometimes even small streams is the 'jug and stopwatch' method. To do this, place a small vessel of known capacity below the spring, and time how long it takes for it to fill. Check that all the flow goes into the vessel and make sure that none of it splashes out. With care, this method is the most accurate way of measuring these flows.

As the name implies, a very popular calibrated vessel for this job is the kitchen measuring jug. Try to get a 1-L polythene jug and avoid those made of brittle plastic that will crack and spoil the measurements. Select a short, fat jug rather than a tall, thin one, as there may not be much space for the jug to fit under the falling water. Accurate measurements need the jug to be upright, so tall ones may not be suitable. Make the measuring mark easy to read by painting over it or sticking on coloured plastic tape. Electrician's insulation tape is ideal for this purpose.

When the spring is running so fast that it fills the jug in less than five seconds, a bigger container such as a 10-L bucket should be used. Again, try to choose one that is short and fat and made of robust material such as polythene. If the bucket has no calibration marks simply put 10 L in it using the measuring jug and mark the 10-L level with paint or tape. Put two marks on opposite sides of the bucket – this will also make it easier to tell if you have kept it level during the measurement.

It may be necessary to modify the spring so that you can catch all the flow in your jug. You may need to divert the flow through a short length of pipe using a temporary dam, as shown in Figure 5.8. Make sure that there are no leaks and all the water goes through the pipe. Ensure that the pipe is high enough and long enough to allow the measuring jug or bucket to fit underneath and stand upright. You may have to dig a shallow hole beneath the pipe to make sure that a bucket will fit. Place a flat stone or concrete slab beneath the pipe for the water to fall on to prevent bed erosion. Springs that are already piped into a cattle drinking trough or a water supply catch pit can be measured as the flow from the overflow pipe. If this is not possible, bail the tank out first so that you can get the jug or bucket below the inlet pipe. When taking measurements on a drinking water supply system, ensure that your hands and the equipment are perfectly clean, and take care not to knock any soil or other debris into the catch pit (see Appendix I).

Sometimes it is not possible to modify a spring in this way, and here you will have to use all your ingenuity for improvisation. If a jug will not fit in the space, cut down a plastic bottle or use a polythene bag to catch the flow. Try to intercept all the flow for at least 30 seconds to achieve a reasonably accurate measurement. Then simply pour the water into your measuring jug and divide the volume by the time taken to collect it. Examples of modifying methods to measure flows are also shown in Case History 3.

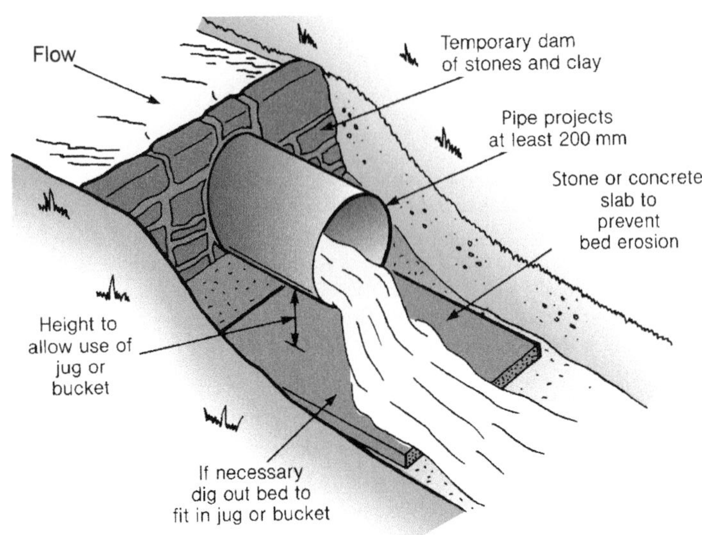

Figure 5.8 *Modify springs so that their flow can be measured with a jug and stopwatch by installing a temporary dam of stones and clay or concrete, through which the pipe projects for at least 200 mm.*

Case History 3 – Measuring Flows with Low Heads

It may be important in an investigation to record a groundwater discharge where the conditions are far from ideal. Two examples are described here from projects I have been involved with over the last few years.

Groundwater flowed into one end of a basement and down a corridor, causing flooding of a large area, and then found its way to a drain some 50 m away. The inflow was dammed off using sand bags, and a plastic drain pipe was used to carry the flow to the drain to keep the rest of the basement dry. At a point in the corridor the floor changes level by about 150 mm, which allowed a specially designed weir tank, using the general principles described in Section 6.4, to be inserted to measure the flow. The shallow depth meant that baffles would be difficult, and so the tank was made with a long wide shape so that the flow would become laminar by the time it reached the 90° v-notch. The bottom of the notch had to be sufficiently above the base of the tank to allow a nappe to form and it was estimated that 40 mm would be sufficient. Box Figure 3.1 shows the dimensions of the tank (a) and the position of the stilling well (b). The calibration of the tank was checked several times using the jug and stopwatch method, and flows compared with the theoretical flows in Table 5.1 showed good agreement.

In a second example, groundwater flowed in 150-mm-diameter salt-glazed earthenware drain pipes beneath the ground, with the depth of water relatively constant and at a depth less than half the diameter of the pipe. A length of the same diameter pipe was obtained and a section removed to provide access.

(a)

(b)

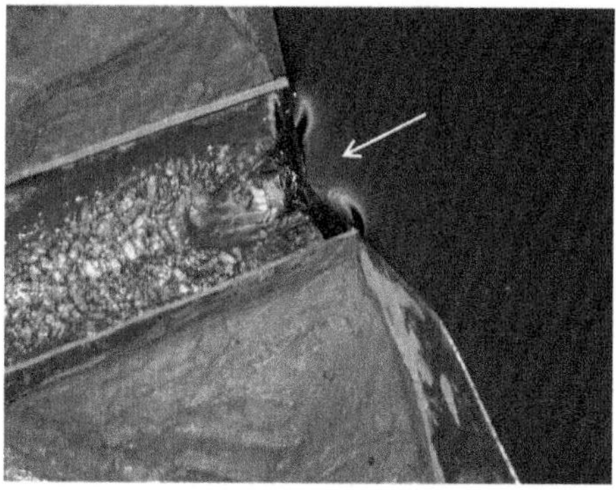

Box Figure 3.1 *The tank (a) was made from 3-mm-thick galvanised-sheet steel bent to shape with welded joints, and a curved section was cut to hold the plastic pipe that carried the discharge. Effective height was limited by the change in floor level of 150 mm. In (b)the arrow shows the direction of flow. The stilling well was made from plastic water fittings and set with its base in a hole through the floor to create sufficient depth to insert a pressure transducer to record the weir tank water levels.*

Table 5.1 Flow over thin-plate weirs in litres per second.

	V-notch weirs			Rectangular weirs			
Head (cm)	¼ 90°	½ 90°	90°	0.6 m width	1.0 m width	1.3 m width	1.6 m width
1	0.005	0.01	0.02	0.8	1.3	1.9	2.1
2	0.2	0.04	0.1	3.1	5.2	6.7	82
3	0.05	0.1	0.2	5.6	9.4	12	15
4	0.1	0.2	0.5	8.7	15	19	23
5	0.2	0.4	0.8	12	20	26	32
6	0.3	0.6	1.3	16	27	35	43
7	0.5	0.9	1.8	20	33	44	54
8	0.7	1.3	2.6	24	41	53	66
9	0.9	1.7	3.4	29	49	63	78
10	1.2	2.2	4.4	34	57	74	91
11	1.5	2.8	5.6	39	66	85	105
12	1.8	3.5	7.0	44	75	97	120
13	2.2	4.3	8.5	50	84	110	135
14	2.7	5.2	10	56	94	122	151
15	3.1	6.1	12	62	104	136	167
16	3.7	7.2	14	68	114	150	185
17	4.3	8.4	16	74	125	163	200
18	5.0	9.6	19	81	136	178	220
19	5.6	11	22	87	148	193	240
20	6.4	12	25	94	159	210	260
21	7.2	14	28	101	171	225	275
22	8.1	16	31	108	183	240	295
23	8.5	17	35	116	196	255	315
24	10	20	39	123	210	275	335
25	11	22	43	131	220	290	360
26	12	24	48	139	235	310	380
27	13	26	52	147	245	325	400
28	15	29	57	154	260	340	420
29	16	31	63	162	275	360	445
30	17	34	68	170	290	380	470

A disc with the same diameter of the pipe was cut from a sheet of 1-mm stainless steel, reduced to about two-thirds the diameter by cutting off the top, and then a 90° v-notch was cut to leave about 30 mm below the notch. The plate was glued in place so that it lay below one end of the access hole, as shown in Box Figure 3.2a.

The depth discharge relationship was determined by measuring the water level using a vernier gauge in the upstream chamber at the same distance from the weir as in the pipe. The discharge was measured by catching the flow in the bucket and timing how long it took to collect 10 L (see Box Figure 3.2b).

(a)

Box Figure 3.2a *View inside pipe shows the stainless-steel 90° v-notch plate with small water flow. The end plate can be seen at the back of the pipe; the hose carrying the mains water supply enters the picture from the right and can be seen on the left at the back of the pipe. Both the weir plate and end plate are fixed with a sealant adhesive.*

(b)

Box Figure 3.2b *General arrangement of the test rig set up on a stack of pallets to provide a firm horizontal platform, with the gaps between the boards modified to hold the pipe. A water supply carried in a hose enters the rear of two holes cut in the top of the pipe. The 10-L bucket was used as the calibrated vessel for flow measurements.*

Box Table 3.1 *Water height and discharge values.*

Water depth	Time for 10 L	Water above	Discharge		
(mm)	(min:sec)	weir (mm)	(L s^{-1})	(m^3 h^{-1})	(m^3 d^{-1})
39.8	10:38	9.8	0.02	0.06	1.35
52.7	01:21	22.7	0.12	0.44	10.61
58.8	00:43	28.8	0.23	0.83	19.88
59.5	00:41	29.5	0.24	0.88	21.12
65.3	00:22	35.3	0.46	1.66	39.72
67.7	00:23	37.7	0.43	1.57	37.57
74.2	00:14	44.2	0.71	2.54	61.02
79.7	00:11	49.7	0.89	3.19	76.60
85.6	00:09	55.6	1.17	4.21	100.93
95.0	00:05	65.0	1.96	7.06	169.41

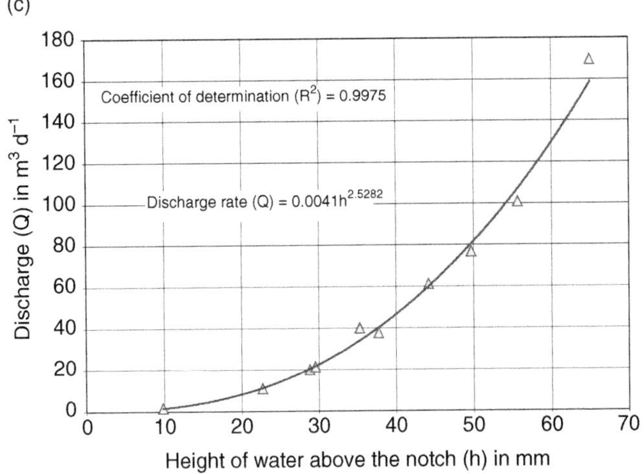

Box Figure 3.2c *Data from lower flows were plotted on the graph and different trend lines were tried. The best fit was the power trend line with an R^2 relationship of 0.9975, which gave the formula $Q = 0.0041h^{2.5282}$ m^3 d^{-1}.*

The data (see Box Table 3.1) were then plotted in Excel and a relationship between the water height (h) and the discharge (Q) was calculated. Because the pipe has a round cross-section, the flows for the low and high ranges were calculated separately.

(a)

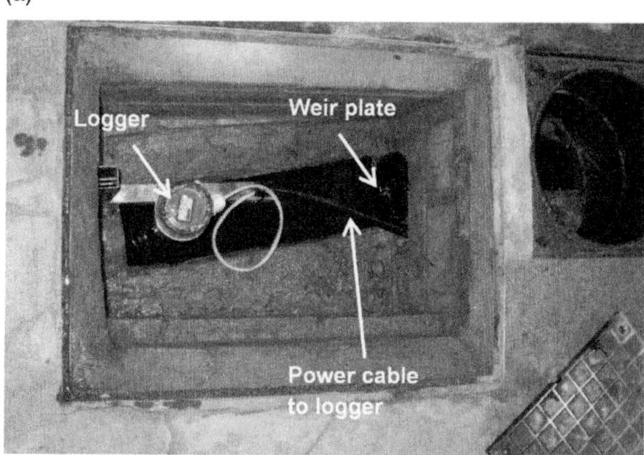

(b)

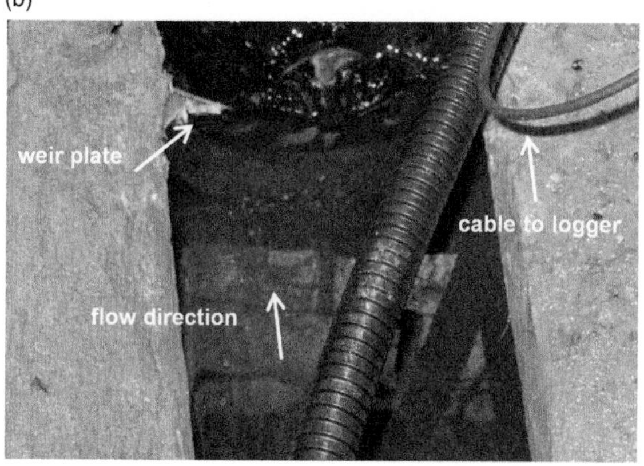

Box Figure 3.3 *(a) Arrangement in the drainpipe. The weir is on the right and the logger is held in a frame fixed to the left-hand wall of the chamber. The two cables are the power supply (black) and the cable to the sensor (white/grey). In (b) the weir plate can be seen with water flowing over it in a nappe.*

Box Figure 3.2c shows the graph for the low range and the formula for calculating the flow. The arrangements in the actual drain pipe are shown in Box Figure 3.3a and b.

Remember to keep the jug or bucket upright, otherwise it will be difficult to decide when the water level has reached the mark. Ensure that the stopwatch is started at the same instant that the first drop of water falls into the bucket, and stop it as soon as the mark is reached. Repeat the measurement at least three times, and take the average as the correct flow. If there is a large discrepancy between the three readings, taking a few more readings will help reduce errors (see Figure 5.9). Experience will soon tell you how many readings to take. Remember that if a jug fills quickly use a bucket, and if that fills rapidly then you should be using a thin-plate weir as described below.

Location:Hogshead spring
Date: 22 10 15 (10.30 am)

Test number	Time to fill 1-litre jug (s)
1	4.95
2	6.04
3	5.55
4	5.73
5	5.48

$$\text{Average} = \frac{27.75}{5} = 5.55 \text{ seconds}$$

$$\text{flow} = 0.18 \text{ litres/s}$$
$$= 15.55 \text{ m}^3\text{/day}$$

Figure 5.9 *Example of a field record of spring-flow measurements. Time taken for the jug to fill to the 1-L mark is measured using a digital stopwatch. Measurements were repeated five times because of the apparent discrepancy between the first two readings. Average time is calculated and the reciprocal taken to give average spring-flow in litres per second.*

5.4 Stream-Flow Measurement

Government agencies in most countries make routine flow measurements on major rivers. Although most attention is given to the major rivers, smaller rivers and streams may have short-term records or a few 'spot' readings where one-off measurements have been made using current meter methods. Records may be expressed as average daily flows, total daily flows or instantaneous flows, all of which are slightly different. It is important to understand these different ways of expressing flows, which are explained in Figure 5.10. The graph represents the flow of a stream as measured continuously at a gauging station. During the 24-hour day represented by the shaded section, the *minimum* flow was 0.2 cumecs and the *maximum* flow was 0.8 cumecs (cumec is an abbreviation of cubic metres per second). The *total daily flow* is represented by the shaded area on the graph, which in this case is 0.48 cumec days. This value is easier to understand if it is converted to 41 472 m^3 d^{-1} by multiplying by the number of seconds in a day (86 400 seconds). The shaded area under the curve also provides the *average daily flow*, which is the arithmetic mean. The value in this instance is 0.48 cumecs or 41 472 m^3 d^{-1}. During the late morning, a spot gauging was taken by current meter (see Section 5.4.1), and this produced an *instantaneous flow* value of 0.35 cumecs.

Stream and river flows are calculated either from measurements of water velocities and the channel cross-sectional area, or by installing a weir. The methods described in the following sections are not suitable for floods or large rivers, but nevertheless will meet the needs of most hydrogeologists.

5.4.1 Velocity–area method

Flow velocities are best measured using a current meter. These are devices that consist of an impeller, which, when placed into a stream, rotates in proportion to the speed of the water passing

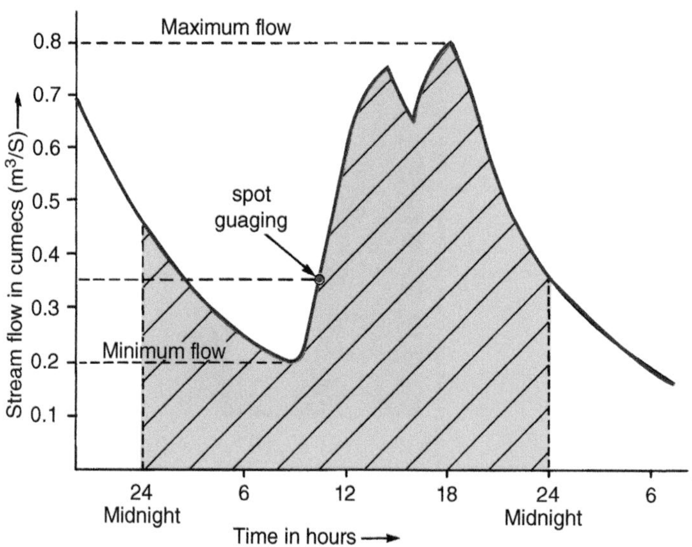

Figure 5.10 *Different ways that stream flow can be expressed. See text for details.*

by it (Figure 5.11a). There are four main types of current meter. The cup type (Figure 5.11b) has an impeller made up of six small cups that rotate on a vertical shaft. The rotor type (Figure 5.11c) has an impeller that looks a bit like the propeller on an outboard motor. In both types the rate of rotation of the impeller is recorded by an electrically operated counter and can be converted to a velocity using a calibration chart or into a direct reading, depending on the type of counter that is used. A third type of current meter is acoustic and uses ultrasonic Doppler technology for highly accurate velocity measurements (Figure 5.11d). It measures the frequency of the transmitted beam reflected off particles in the water, with the frequency shift used to calculate the velocity. These instruments are maintenance free and have integrated depth measurements and graphical step-by-step user guide to make discharge calculations. A fourth type consists of an electromagnetic current meter that produces a voltage proportional to the velocity and has no moving parts (Figure 5.11e). Advantage of this type of current meter are direct analogue reading of velocity, it automatically calculates discharge and it is maintenance free. It can be used in low-flow environments, and is unaffected by large amounts of organic matter in the water. Generally, because current meters are expensive to buy and maintain, they will only be available to hydrogeologists working in organisations such as water companies, university departments and large consulting firms.

All current meters need some maintenance, if only cleaning after they have been used. Those that have moving parts need routine calibration (usually once a year) carried out by the company you bought them from. Make sure that you follow the manufacturer's recommended frequency for oiling and clean them every time they have been used. Electromagnetic meters only need to be cleaned, although they will require annual calibration.

Current meters are used by suspending them in the stream, pointing in an upstream direction. For deep streams this must be done from a bridge or a specially constructed cableway. In most cases, you will probably be able to stand in the stream wearing thigh-waders when you make these measurements. In such cases, the current meter is attached to a special pole, graduated in centimetres, which allows the instrument to be positioned into the river to a known depth. The rate at which the

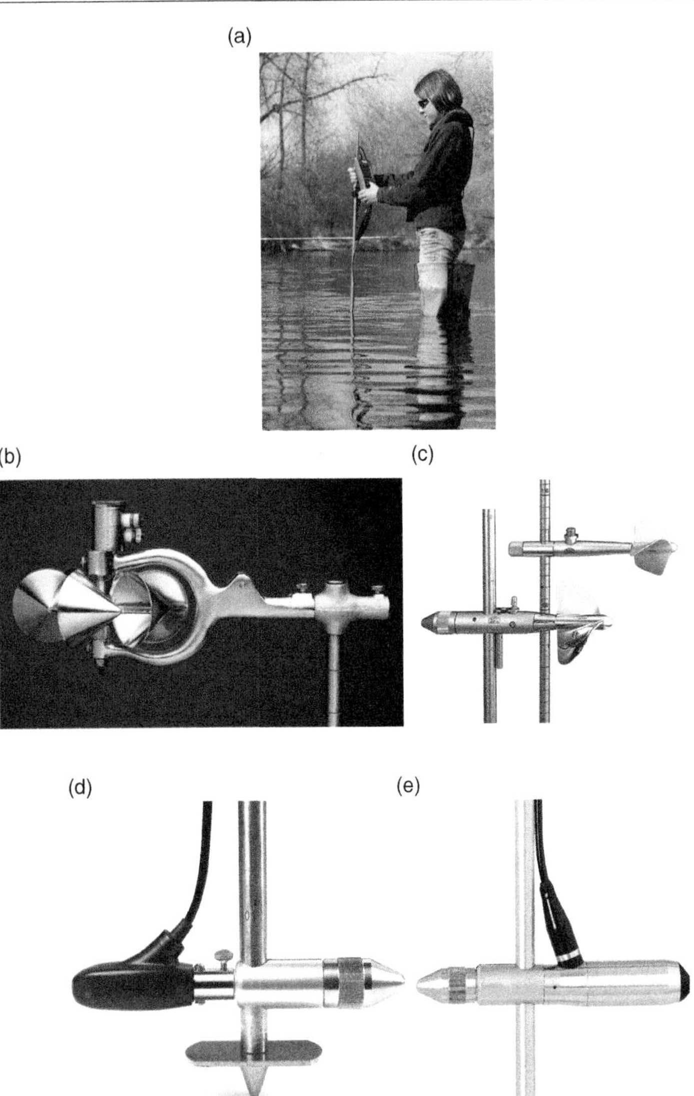

Figure 5.11 *Four types of current meter in common use. In (a) the operator is using one of the current meters to gauge stream flow. The cup-type (b) has an impeller made up of six small cups that rotate on a horizontal wheel. The rotor type (c) has an impeller that looks like the propeller on an outboard motor. In both types, rate of rotation of the impeller is recorded by an electrically operated counter and is converted to velocity using a calibration chart. These instruments should be periodically recalibrated by a specialist firm. The acoustic current meter (d) uses ultrasonic Doppler technology to ascertain water flow velocity. The electromagnetic current meter (e) produces voltage proportional to velocity. (Reproduced with permission from Gurley Precision Instruments, Inc. and of OTT HydroMet / https://www.ott.com/ / last accessed under March 24, 2023).*

impeller rotates is recorded by an electronic counter suspended around the operator's neck and attached to the current meter by a thin cable.

Choose a straight stretch of stream with a fairly constant depth for your gauging and avoid places where the flow is impeded by cobbles, boulders, bridge supports, and so on. Figure 5.12 shows the steps to take when making a current meter gauging. First, stretch a surveyor's tape across the stream at right angles to the direction of flow. You will use this to locate each station where you take the velocity readings at regular intervals across the stream, spaced so that there are between 10 and 20 stations altogether. Choose equal spacings of an easy size for the calculation, such as a metre or half a metre, and give each station a reference number to identify it. The gauger faces upstream and stands to one side of the instrument, so that turbulence around his or her legs does not affect the readings, and not in front or behind the instrument.

Figures 5.13 and 5.14 give details of how to record field measurements and also how to calculate the flow. (Appendix I includes safety precautions that must be observed when making this type of field measurement.)

In theory, the average velocity occurs at 0.6 of the depth of the stream measured downward from the surface, and this is the position that all textbooks tell you to take your reading. For most practical purposes, however, it is sufficiently accurate to take a measurement at half the depth and then use a correction factor (see Figure 5.13). After all, when you are standing in a river, it is much easier to halve a depth in your head than multiply it by 0.6! And the results will be sufficiently accurate.

First measure the depth of water using the graduated pole and set the current meter at half this value. Place the bottom of the rod on the streambed, so that it stands vertically and at the precise position on the surveyor's tape. The current meter should point straight upstream and you should stand downstream and slightly to one side so that the measurement is not affected as the water flows around your legs. Allow the rotation of the impeller to settle down for a few seconds before starting

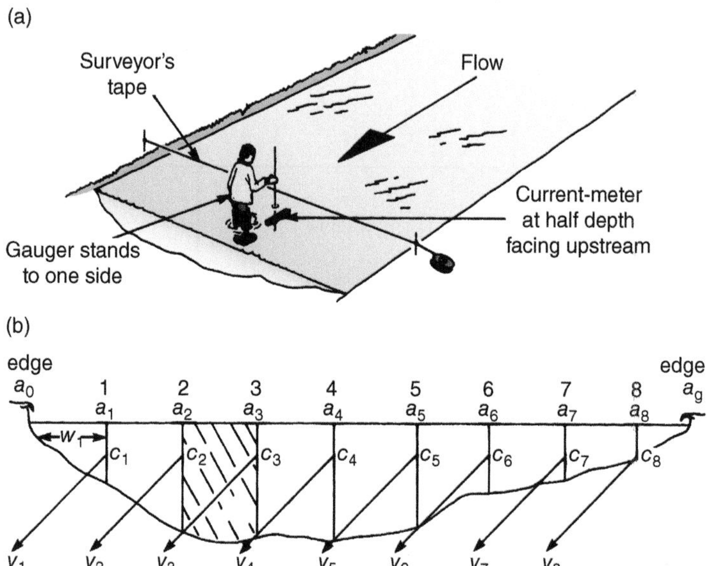

Figure 5.12 *Stream flow is often measured by means of a current-meter gauging (a), where stream-flow velocities are measured at regular intervals across the stream, as shown in (b).*

(n) Section number	(a) Tape reading (m)	(w) Section width (m)	(c) Depth reading (m)	(d) Mean depth of section (m)	(A) Area section (m^2)	(r) Current-meter reading at 0·5 depth: Field (rpm)	(x) Velocity (from tables) (m/s)	(v) Mean velocity of section (m/s)	(Q) Flow for section (m^3/s)
Edge	a_0		Zero			Zero	Zero		
		$w_1 = a_1 - a_0$		$d_1 = \frac{2 \times c_1}{3}$	$A_1 = w_1 \times d_1$			$v_1 = \frac{2 \times x_1}{3}$	$Q_1 = A_1 \times v_1$
1	a_1		c_1			r_1	x_1		
		$w_2 = a_2 - a_1$		$d_2 = \frac{c_1 + c_2}{2}$	$A_2 = w_2 \times d_2$			$v_2 = \frac{x_1 + x_2}{2}$	$Q_2 = A_2 \times v_2$
2			c_2			r_2	x_2		
		$w_3 = a_{3-a_2}$		$d_3 = \frac{c_2 + c_3}{2}$	$A_3 = w_3 \times d_3$			$v_3 = \frac{x_2 + x_3}{2}$	$Q_3 = A_3 \times v_3$
3	a_3		c_3			r_3	x_3		
		$w_4 = a_4 - a_3$		$d_4 = \frac{c_3 + c_4}{2}$	$A_4 = w_4 \times d_4$			$v_3 = \frac{x_3 + x_4}{2}$	$Q_4 = A_4 \times v_4$
etc.	etc.		etc.			etc.	etc.		
		etc.		etc.	etc.			etc.	etc.
		$w_9 = a_9 - a_8$		$d_9 = \frac{2 \times c_9}{3}$	$A_9 = w_9 \times d_9$			$v_9 = \frac{2 \times x_8}{3}$	$Q_9 = A_9 \times v_9$
Edge	a_9		Zero			Zero	Zero		

Calculation of stream flow	
(1) Enter field readings *n*, *a*, *c* and *r* and obtain *x* from tables supplied with the current meter. (2) Calculate *w*, *d*, *A*, *v* and *Q*. (3) Sum values of *Q* (4) Correction for 0.5 depth. Stream flow = $0.95 \times Q$ m³/s = $950 \times Q$ litres/sec	Total Flow = *Q*

Figure 5.13 *Field measurements to record during a stream-flow gauging, and how to use them to calculate flow. All symbols have the same meaning as Figure 5.12. Record all depths and distances in metres. Some instruments automatically convert number of revolutions into velocity; otherwise use tables supplied with the instrument.*

Date: 11TH JULY 2016 Stream: Farbeck: (U/S Farby Road Bridge)
Time: START : 12·45 FINISH : 13·05

n Section number	a Tape reading (m)	w Section width (m)	c Depth reading (m)	d Mean depth of section (m)	A Area of section (m^2)	r Current-meter reading at 0.5 depth: Field revs in 50 secs	x Current-meter reading at 0.5 depth: Velocity (m/s)	v Mean velocity of section (m/s)	Q Flow for section (m^3/s)
1	0.75		edge		*	—			
1–2		0.25		0.167	0.0418			—	—
2	1.0		0.25			no detectable flow	—		
2–3		0.5		0.280	0.1400			0.229	0.0416
3	1.5		0.31			278	0.343		
3–4		0.5		0.335	0.1675			0.354	0.0593
4	2.0		0.36			298	0.365		
4–5		0.5		0.365	0.1825			0.330	0.0602
5	2.5		0.37			235	0.295		
5–6		0.5		0.395	0.1975			0.282	0.0557
6	3.0		0.42			211	0.269		
6–7		0.5		0.415	0.2075			0.284	0.0589
7	3.5		0.41			237	0.298		
7–8		0.5		0.385	0.1925			0.266	0.0512
8	4.0		0.36			180	0.234		
8–9		0.5		0.405	0.2025			0.236	0.0478
9	4.5		0.45			183	0.238		
9–10		0.5		0.435	0.2175			0.223	0.0485
10	5.0		0.42			156	0.208		
10–11		0.5		0.435	0.2175			0.169	0.0368
11	5.5		0.45			87	0.131		
11–12		0.5		0.425	0.2125			0.152	0.0323
12	6.0		0.40			124	0.172		
12–13		0.5		0.340	0.1700			0.151	0.0257
13	6.5		0.28			86	0.130		
13–14		0.5		0.440	0.2200			0.087	0.0342
14	7.0		0.16			no detectable flow	—		
14–15		0.4		0.107	0.0428			—	
15	7.4		edge		*	—			
								Total flow	0.5522

* Note : As there was no detectable flow at 2 and 14 section areas have been added.

Correct for 0·5 depth
flow = 0·5522 x 0·95
= 0·525 m^3/s or 525 litres/sec.

Figure 5.14 *This example shows how field data have been used to calculate stream flow using the method given in Figure 5.13. In this case, the current meter only recorded revolutions and not velocities. Velocity measurements were taken at half the depth and so, once total flow had been calculated, it was corrected by multiplying by 0.95.*

to count. Many modern instruments have a timing device in the counter; otherwise use a stopwatch and record the number of counts over a one-minute period. Repeat this procedure at each station.

Record the reference number for the station, the distance from the bank, the depth of the water, and water velocity or the number of rotations in your notebook for each measuring station. The calibration data sheet for the current meter is used to convert the number of rotations per minute into a velocity. The average velocity for each segment is calculated and multiplied by the cross-sectional area to give the flow for each segment. The sum of the segment flows is the stream flow. Figure 5.13 shows how these field measurements are recorded and used to calculate the stream flow, with an example being given in Figure 5.14. All the symbols have the same meaning as in Figure 5.12. Take care to record all depths and distances in *metres*, not centimetres. Most instruments automatically convert the number of revolutions of the impeller into a velocity; otherwise use the tables supplied with the instrument.

You can make a rough and ready flow estimate for small streams using a crude form of the velocity–area method. This technique is useful when you are making an initial reconnaissance of the study area and want to get some idea of the general order of stream flows. First, choose a length of stream that runs straight for 4 or 5 m. Use a steel tape to measure both the width and depth of the channel at right angles to the direction of the flow. Multiply these values to get the cross-sectional area of the stream. If the streambed is irregular, measure the depth at various distances from the bank and calculate an average depth. The water velocity can be estimated by timing a piece of stick, or other suitable small float, over a measured distance of a metre or two. If the water is flowing rapidly it may be prudent to increase this distance to 5 or even 10 m. You should take several readings, with the float being placed at different distances across the stream, and then take the average.

The flow is calculated by multiplying the velocity measured in metres per second by the cross-sectional area measured in square metres. This will give the answer in cubic metres per second, and to convert to litres per second multiply by 1000. This method will overestimate flows, as the surface velocity is significantly greater than the average velocity. *It is important, therefore, to correct the estimated value by multiplying it by 0.75.*

5.4.2 Thin-plate weirs

The most accurate method of measuring the flow of a stream is to use a thin-plate weir. All weirs work by restricting the size of the stream channel. This causes the water to pile up on the upstream side, before passing over the weir as a jet (or *nappe*). The flow over the weir is proportional to the height of water above the weir on the upstream side. It is relatively easy to record these water levels and then read the flow from standard tables.

Thin-plate weirs have either a v-notch or a rectangular notch. The dimensions and method of operation are described in official standards such as the British Standard ISO 1438:2017. The specification given in such standards can mean accurate flow measurements to within 1% in favourable conditions.

You do not need to buy a special weir plate, as they are quite easy to make. Provided that you follow the general guidelines given here for their construction and installation, it is possible to achieve flow measurements with an accuracy better than ± 10%. Sheet metal or plywood can be used to make a weir plate. It is important to cut the angles as accurately as possible to make sure that all edges are sharp and straight and that the upstream face is smooth. Figure 5.15 shows the general features of v-notch weir plates. The weir should have a lip of between 1 mm and 2 mm and the downstream face should slope away from the lip at an angle of at least 60° in the case of the v-notch weir. For rectangular weirs this angle must be at least 45°. The range of flows each weir can measure is indicated in Tables 5.1 and 5.2.

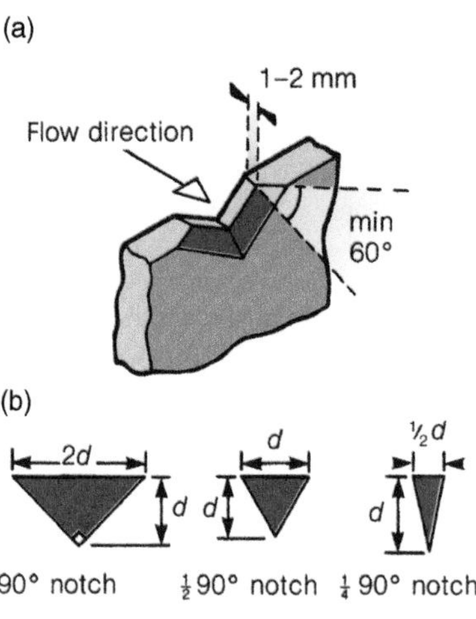

Figure 5.15 *It is important that a v-notch weir is made to the correct specification for accurate measurements to be achieved (see main text). The general arrangements shown in (a) apply to the three types of v-notch shown in (b).*

Table 5.2 *Maximum flows for accurate measurement with different methods.*

Method of measurement	Maximum flow for accurate measurement (L s^{-1})
Jug and stopwatch	
1-L jug	0.25
10-L bucket	3
Weir plates	
¼ 90° v-notch	17
¼ 90° v-notch	34
½ 90° v-notch	68
Rectangular weirs	
0.6 m width	170
1.0 m width	290
1.3 m width	380
1.6 m width	470

There are three types of v-notch weir, each with a slightly different angled notch. These are a 90° notch, which has its width across the top equal to twice the depth, a ½ 90° notch, with the width across the top equal to the depth, and a ¼ 90° notch, which has its width across the top equal to half the depth. If these dimensions are used, it is a relatively simple job to construct the weir by cutting an appropriate isosceles triangle from the plate. The angles of the ½ 90° and ¼ 90° v-notch weirs are not 45° and 22.5° as might be expected, and get their name from the fact that they measure half and one-quarter the flow rate of a 90° v-notch.

5.4.3 Installation and operation of thin-plate weirs

Select a straight section of stream channel at least 3 m long and install the weir at the downstream end. Installing these weir plates is not an easy job, often requiring two people, and is likely to make you both wet and dirty. Do not take short cuts, as poor workmanship will mean that the weir will be washed out in a few days or even less. The installation is illustrated in Figure 5.16.

Values have been calculated from standard formulae. Flow over a 90° v-notch, weir is given by the formula $Q = 1.342h^{2.48}$, where Q is the flow (in $m^3\ s^{-1}$) and h is the head (in m). Flow over a $\frac{1}{2}$ and $\frac{1}{4}$ 90° v-notch weir is half that over a 90° v-notch weir, and $\frac{1}{4}$ 90° notch is a quarter of that over a 90° v-notch, respectively. Flow per 1-m length of rectangular weir is given by the formula $Q = 1.83(1 - 0.2h)h^{1.5}$, where Q and h have the same meaning and units as before.

It is easier to install a weir plate if you use sandbags to divert the stream or reduce the flow while you work. Dig out a shallow trench in the bed and banks at right angles to the flow to take the plate. Work quickly as the trench may collapse! Use a builder's level to make sure that the weir plate is upright and support it if needed with stakes driven into the bank. The level of the crest of a rectangular weir or the apex of a v-notch weir should be set at a height above the stream bed to give an adequate fall on the downstream side. At maximum flows this must be at least 75 mm above the water level on the downstream side. It is vital to make a good seal in both bed and banks; otherwise leakage will cause erosion and the eventual collapse of the structure. Seal the edges with clay by pushing it in with your fingers. Prevent bed erosion, which will also cause collapse, by placing stones or a concrete slab below the weir to take the impact of the falling water. Remove any temporary sandbag dam carefully so as not to cause a surge of water that may wash out the weir. Install the gauge board

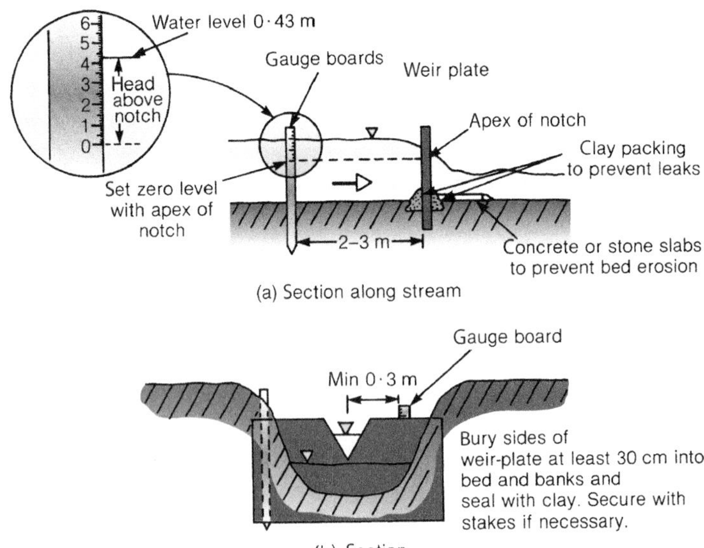

Figure 5.16 *Installation of a weir plate.*

on a vertical post at the appropriate upstream distance and position it to one side of the channel, not in the centre. Use a builder's level to ensure that the post is vertical. The gauge board should be accurately marked off in centimetre graduations. A wooden ruler can be used as a ready-made gauge board, with alternate marks painted in a distinctive colour to help to read the water level to the nearest centimetre. Use the builder's level to set the zero of the gauge board at the same level as the crest of the rectangular weir or the apex of the v-notch. The flow of water over a weir is related to the depth of water above the weir crest by standard formulae. The depth of water is measured using a gauge board located at the correct distance upstream of the weir plate, as shown in Figure 5.16. The relationship between water level and flows over various types of thin-plate weir is given in Table 5.1.

Thin plate weirs tend to silt up on the upstream side and need to be cleaned out so that the water is deep enough for the weir to work. For a rectangular weir, the minimum depth of water required on the upstream side is 60 cm. A 30-cm-high v-notch will need a depth of 45 cm, and a 15-cm-high notch will require 30 cm of water. Clean the section out *before* the flow measurement is made and remove debris such as twigs, branches, waterweeds, polythene bags and other trash from the weir. Do not take the reading until stable conditions have been re-established. This may take several minutes, so be patient. Normally one reading would be taken each day and be used to build up a record of flows.

5.5 Stage–Discharge Relationships

When it is not possible to install a thin-plate weir, flows can be measured by relating stream levels (*stage*) to a series of current meter gaugings (*discharge*). Select a straight length of stream where flows are controlled by a natural feature such as a rock outcrop. If this is not possible, an artificial control can be installed, such as a length of 50-mm-diameter pipe secured to the streambed by stakes and extending between the banks at right angles to the flow. Install the gauge board on a wooden stake set vertically on the side of the channel about 4 or 5 m upstream of the control. Current-meter gaugings are made as described above, with the gauge board being read both before and afterwards in case the stream level changed while you took the measurement. If there is a difference, relate the flow to the average level during the gauging period.

These measurements will build up a record of the relationship between stream level and flows so that a stage–discharge graph can be drawn, as shown in Figure 5.17. Simple relationships can be seen from a plot on a natural scale (a), but more complex relationships are best understood if a log–log graph is plotted (b), as the relationship then plots as a straight line. This graph can be used to estimate flows from gauge-board readings without a current meter measurement being carried out.

For a continuous record, install a water level recorder or data logger system set to the same datum as the gauge board. The instrument will need to be in a box to protect it from both the elements and vandals and with the instrument installed in a temporary stilling well, as shown in Figure 5.18. In this example, a stilling well has been made using a vertical pipe set on the streambed and secured to the bank with timbers. Holes drilled in the pipe ensure that the water level in the pipe is the same as that of the stream. This arrangement removes small fluctuations in water level caused by waves and protects the instrument. An artificial bed control has been provided by a 50-mm-diameter pipe secured to the streambed with stakes several metres downstream of the recorder site and set at right angles to the flow. A gauge board has been installed next to the stilling well so that the recording instrument can be calibrated against spot readings. The water level record collected in this way can be converted into flows using the stage–discharge graph. If you plot the stage–discharge data in Microsoft Excel, you can easily obtain the equation for the best-fit curve and then use this to convert the water level record into flows.

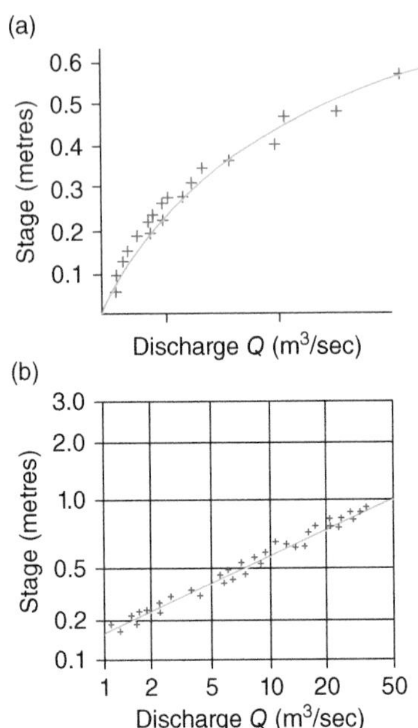

Figure 5.17 *A stage–discharge graph shows the relationship between stream flow and stream-water level at a particular point.*

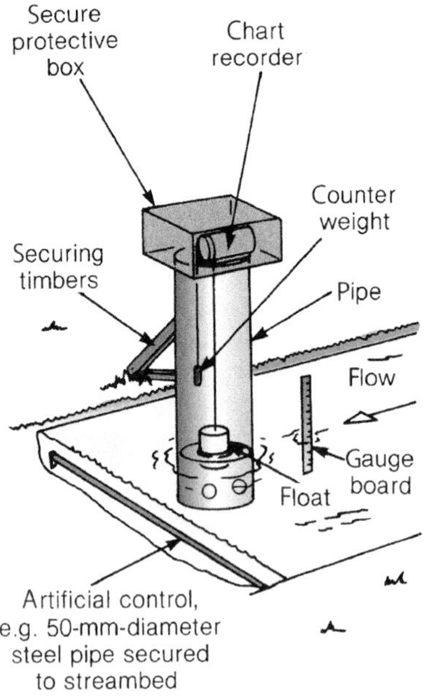

Figure 5.18 *A continuous record of stream level is made by a temporary installation of a water level recorder or data logger system.*

5.6 Choosing the Best Method

The ideal measurement method for spring or stream flows will depend on local circumstances and must be suitable for the range of flows you are likely to encounter. In many groundwater studies the low flow range will be most important, and so you should select the most appropriate method of field measurement for the flows at this end of the range. Table 5.2 will give you an idea of the maximum flows that can be measured accurately with each method.

Look at the stream channel dimensions to decide which method is best to measure the flows. A wide, shallow channel with a small downstream slope, for example, may be unsuitable for using a thin-plate weir, and you may have to resort to the jug and stopwatch method after building the sort of installation shown in Figure 5.8. One word of warning before you get out your spade and start to dam up a stream. In many countries it is necessary to obtain the permission of the environmental regulator or local authority to install equipment in a watercourse, and you should certainly talk to local landowners. Quite often the formal aspects will be minimal and you may only have to let the local office know about your proposals. This may well have hidden bonuses, because the resident hydrologist will probably be very pleased to provide advice on the best way of tackling the problem.

5.7 Processing Flow Data

Plotting flow data against time is an excellent way of comparing measurements from different sites and, if you include rainfall readings, it is possible to tell how quickly flows respond to rainfall events. Prepare the data by converting flows to the same units, and then select a time scale large enough for all the measurements to be plotted against the time they were taken. You should expect a short time lag of only a few hours before flows of streams and rivers start to increase after heavy rain. The precise length of this response depends upon many factors, such as catchment size, slope, and geology, the amount of rainfall over the recent past few days, the intensity of the rain, and the direction in which the storm was travelling across the catchment. Groundwater levels, on the other hand, respond much more slowly, and variations in spring flows generally follow a similar pattern as the groundwater levels. Springs that drain deep aquifers may change very little throughout the year, and you are unlikely to see the effect of an individual rainfall event. The flow of springs fed by shallow aquifers varies much more, and this factor can be used to identify each type of spring.

The example shown in Figure 5.19 is based on field data, and shows how marked these differences can be. In Figure 5.19, the rapid response of Pott's Brook to rainfall is obvious, as is the different response of the two springs. The Croft Farm Spring flow increases with rainfall, suggesting that it is the discharge from a shallow aquifer or even a land-drain. The flow of the High Farm Spring changes very little, indicating that it is the discharge from a separate, deeper aquifer. Rainfall at the nearby Gately sewage works is also shown.

Stream-flow data can be examined to identify the groundwater component using a technique called *baseflow separation*. Total stream flow is made up of a number of components, each of which behaves differently. The two major components are surface runoff and groundwater drainage. The latter is made up of spring flows and direct discharges into the stream, and is referred to as *baseflow*.

A graphical technique is straightforward to use and involves plotting flow data on semi-logarithmic graph paper, with flow on the vertical axis on a logarithmic scale and time along the bottom on a natural scale, as shown in Figure 5.20. The decline in groundwater flows (or *baseflow recession*) approximates to a straight line with a constant gradient. Once the data are plotted, examine the slope of the graph after single-peak storms. The decline in total stream flow will follow the baseflow recession line at the later stages of this flow reduction. Compare a number of single storm events. This is important, as the baseflow recession will be a constant and so the various straight lines should be parallel. Once the slope of this line has been identified, you can use it to draw in the baseflow in those parts of your records where a straight line is not so obvious. In this way it is possible

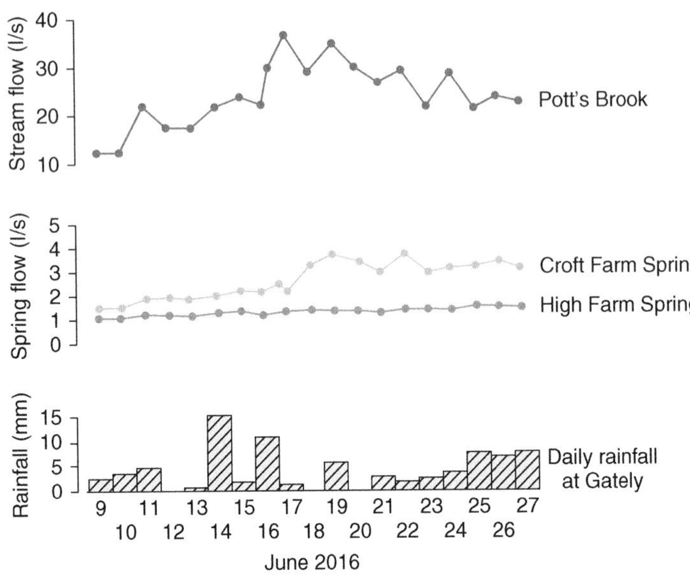

Figure 5.19 *Relationship between spring flow and rainfall is easily discovered if records are plotted on the same time scale.*

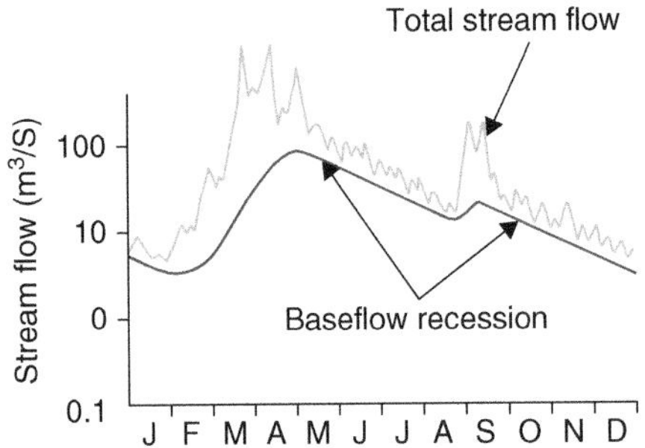

Figure 5.20 *A base flow graph can be constructed from a semi-logarithmic plot of stream flow on the log-scale against time on the linear scale. Rate of base flow recession is a constant, which produces a series of parallel lines as shown. The groundwater component of stream flow is represented by the area on the graph beneath these straight lines.*

to separate the groundwater component from the other elements of stream flow. The groundwater component of the steam flow is represented by the area on the graph beneath these straight lines.

Spring flows can be examined in the same sort of way. As many 'springs' may really be land-drains or at least have a surface water component, it can be important to identify the groundwater element. This will help you to decide which aquifer unit a spring is draining, and will indicate whether or not it is likely to dry up. Both of these facts are fundamental to a full understanding of the workings of the groundwater system in your study area.

6
PUMPING TESTS

When a pumping test is carried out carefully it can provide data that can give a deep insight into borehole yields, aquifer properties and groundwater flow. Consequently, pumping tests play an important role in hydrogeological field investigations. However, please do not make the common mistake of calling them *pump tests* – strictly speaking, these will only find out if a pump is working properly!

Pumping tests are time consuming and are often costly. It is important, therefore, to plan and carry them out with care to obtain the best quality information for the costs and effort involved. The data produced from pumping tests may not always be easy to interpret, usually because the geological conditions and therefore the groundwater flow systems are complex. This chapter explains how to plan and carry out the usual types of pumping test and provides an introduction on how to interpret the data obtained from them.

Tables B.1 to B.7 contain conversion factors for different units used in hydrogeology. I use metric units with time in hours and days. I know that others in the UK use metric units with time in seconds and in other countries different units are used.

6.1 What Is a Pumping Test?

A pumping test is a means of investigating how easily water flows through the ground into a well. It consists of pumping in a controlled way at predetermined rates and measuring the resulting effects on water levels in both the pumping well and observation boreholes. Other measurements include spring and stream flows and groundwater chemistry.

Pumping tests can be divided into five types as follows.

1. *Proving tests*. Used to establish the yield of a new well or check on the performance of an existing one. They may only take eight hours for small water supplies or up to a week or so for major abstractions.
2. *Step tests*. Normally completed in one day, these tests involve pumping at different rates for three or four equal periods. The relationship between pumping rate and drawdown is used to define the hydraulic characteristics of the well, allowing the most efficient pump to be selected. The test normally includes monitoring the recovery of water levels after the pump has been turned off.
3. *Impact tests*. Used to establish the impact of a new abstraction on other water users and the water environment in general. The borehole is pumped at a constant rate for several days or longer and this is followed by a recovery test. Water levels are monitored in surrounding wells, and spring and stream flows may be measured.
4. *Aquifer tests*. A constant rate and recovery test designed to provide information on an aquifer's hydraulic properties and involves measurements in a number of observation wells over a period of a few days to several weeks. Again, water levels are monitored in surrounding wells, and spring and stream flows may be measured.

Field Hydrogeology, Fifth Edition. Rick Brassington.

5. *Tests on single boreholes*. A number of short-term tests lasting only an hour or two provide limited information on the hydraulic conductivity. They involve monitoring changes in water level in piezometers caused by water being removed or added. Such tests are termed *bailer* or *rising head*, and *injection* or *falling head* tests, respectively.

Several tests may be carried out in a continuous pumping test programme that might comprise setting up the site and calibrating the equipment on day 1, performing a step test on day 2, followed by overnight recovery, and on day 3 starting a constant-rate test of several days' duration, with a recovery test at the end.

6.2 Planning a Pumping Test

Pre-planning is essential to ensure that you monitor everything that is required to meet the objectives. Table 6.1 provides a checklist to help you complete this planning process. Once you have decided on the form of the test, you should assemble the equipment you will need, using Table 6.2 as a guide. Make sure that all the equipment is in working order before taking it to site and do not forget a supply of the spares and, of course, your toolkit (Table 4.2). Finally, check out the safety implications for all aspects of the test, so that you and your helpers are not exposed to danger during the test programme.

Table 6.1 *Checklist for pumping test planning.*

Length of test	Use Table 6.5 to decide on duration of the test.
Pumping rate	Make sure that the hourly or daily rate is greater than the proposed operational pumping.
Discharge	Ideally use a weir tank. Otherwise use a flow meter with check readings to ensure accuracy. Choose discharge location to avoid recirculation and use pipes or line ditch to carry water to discharge point. Settle suspended material before discharge. Check to see if permission from land owner and consent from regulator are required.
Water levels	Is there a dip tube or can one be installed? What is the likely depth of the pumping water level? Will the dipper/probe cable be long enough?
Observation wells	Needed to determine aquifer properties and identify impacts.
	Depth and casing details needed to confirm aquifer.
	Can other nearby abstraction wells be shut down or pumped at a constant rate throughout the test period?
Other observations	May include rainfall, barometric pressure, spring flows, stream/river flows, chemical parameters using in-line probes or sampling. Before the start of the test make sure that all sites are suitable and modify where necessary. Check that you have the necessary permissions.
Pre-test and post-test monitoring	Decide on period needed to monitor observation wells and parameters listed above to establish trends and possible effects from other causes (e.g. other abstractions). Decide on length of post-test monitoring period. May be dictated by the environmental regulator.
Safety	Look at all the places where measurements are to be taken during the test to ensure that there are no hidden dangers, such as confined spaces at the wellhead, unprotected moving parts on pumps, unprotected electrical switchgear and water treatment chemicals. If the test is to go on overnight, arrange accommodation and adequate lighting for night-time readings. Write down a safety plan.

Table 6.2 Checklist for pumping test equipment.

Pump discharge	Weir tank, or flow meter with possible check readings using a jug and stopwatch or bucket and stopwatch.
Water levels	Dipper: ensure there are enough for all measurement points, plus a spare. Make sure that tapes are long enough to reach pumping-water level. Test each dipper in the office and take spare batteries. Pressure transducer/data logger: ensure that transducer range will cover expected changes in water level. Check equipment in the office and at the wellhead after it is installed.
Pumps	Install the pump in the test well at least 24 hours before the start of the test and make sure it works.
Water quality	Use information in Tables 7.3 and 7.4 to assemble the sampling and monitoring equipment.
Springs and streams	Jug and/or bucket and stopwatch, weir plates, current meter.
Other measurements	Rain gauge, recording barometer or pressure transducer/data logger set for barometric readings, or use local weather station.
General	Notebook, pumping test record sheets on clipboard, semi-logarithmic graph paper, spare pens and pencils, calculator (with spare batteries) or laptop computer (make sure it is fully charged), toolkit, protective clothing, first-aid kit, mobile phone (cell phone).

6.3 Pumps and Pumping

The choice of the right pump is critical to the success of a pumping test. The pump must be capable of matching the maximum pumping rate you want to achieve and also lift the water to the total height (or head) to the discharge point. This total head is made up of the height between the pumping-water level and the ground surface, plus any additional height needed to lift the water to reach the weir tank or discharge point, plus an allowance for the friction losses caused by water flowing through the pipes. Friction losses vary tremendously according to the pumping rate, the pipe diameter, the number of bends and the pipe material. In practice, adding an extra 10–20% to the total head is usually sufficient to select the pump capacity needed.

6.3.1 Surface suction pumps

A surface suction pump can be used for pumping tests where the rest-water level is less than 5 m below ground level. They can be rented from tool-hire companies.

It is always better to have the pump capacity bigger than you need, rather than too small, so if in doubt err on the larger side. In theory, a suction pump can lift water to a height equal to atmospheric pressure (about 10.3 m). In practice, the maximum lift with the most efficient pumps is 7–8 m, although many can only manage only 5–6 m. Despite these restrictions, suction pumps can be used to test most dug wells and even some boreholes. Hire pumps may be powered by petrol or diesel engines. Observe safety precautions and follow the guidelines for fuel transport, storage and refuelling given in Appendix I.

6.3.2 Submersible pumps

Where the water table is deeper than 5 m you can either hire a submersible pump or set up a temporary airlift pumping arrangement. If you hire a submersible pump make sure that the supplier installs it deep enough to allow for drawdown so that it will remain under water, and check that it is working before the supplier leaves the site. Fit a valve on top of the rising main to enable you to adjust the pumping rate. Figure 6.1 shows a typical arrangement for an electrical submersible pump in a borehole. Most submersible pumps are powered by electricity from either a mains supply or a diesel-driven generator. Some pumps are driven directly by a diesel engine using a series of rods that run down through the rising main to the pump. They are more difficult to install than electrical submersibles, and are rarely used in pumping tests for that reason.

6.3.3 Airlift pumping

Airlift pumping uses a supply of compressed air that is forced through a pipe set well below the water level. The air bubbles mix with the water to form an emulsion that is lighter than water and so floats upwards. A continuous air supply keeps the air/water mixture moving quickly up the pipe so

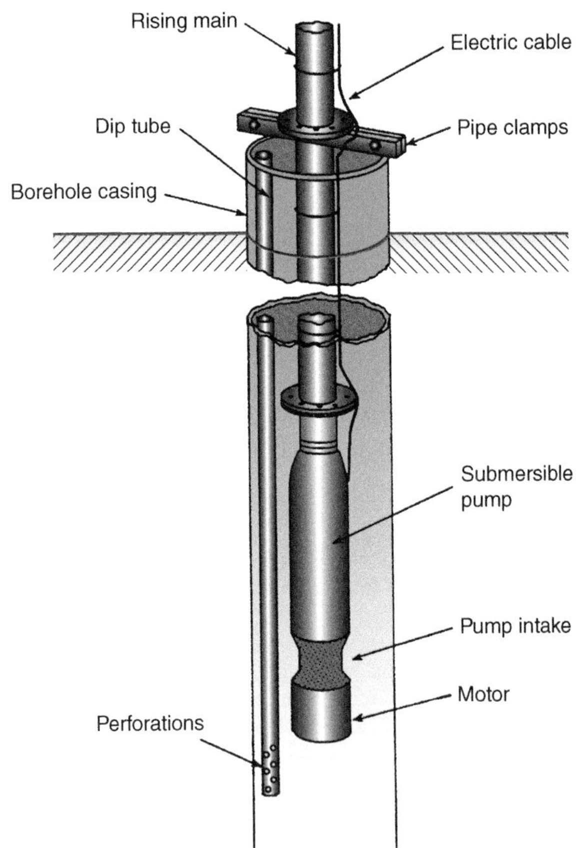

Figure 6.1 *An electrical submersible pump hung on a bolted-flange jointed rising main and suspended in a borehole on pipe clamps that rest on top of the borehole casing. Alternatives include hanging the pump on a flexible rising main that comes in one length to ensure that the pump is at the predetermined depth or where small pumps are involved using a small-diameter plastic pipe, again in one length. The electrical cable is tied at intervals to the main. A dip tube is installed which extends to below the pump intake.*

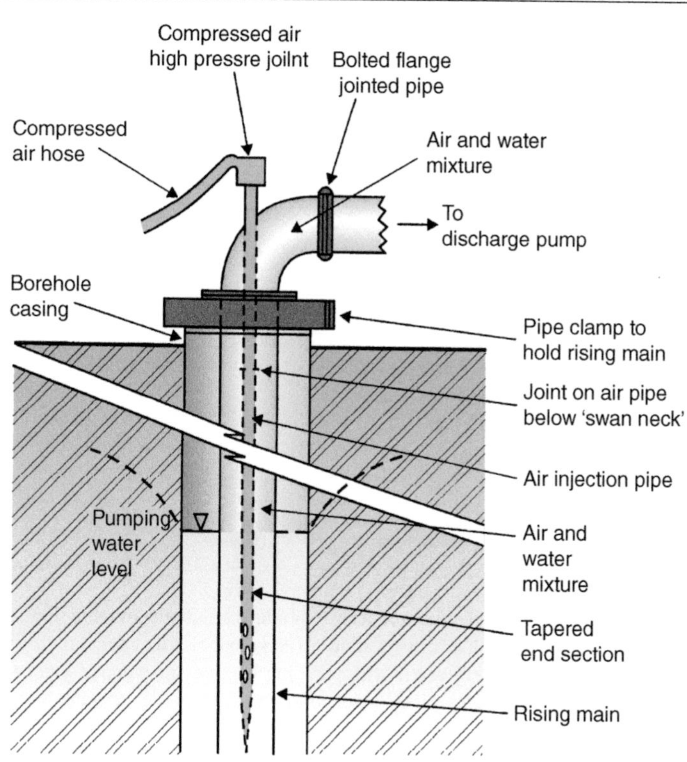

Figure 6.2 *Air-lift pumping uses the sort of equipment shown here.*

that water flows out at the surface. Figure 6.2 shows how to arrange the pipework for an airlift pump. The air is conducted into the rising main through a 'swan-neck' made from a length of steel pipe (with the same diameter as the rising main) bent into a slow 90° angle, as shown. A length of smaller diameter (25–50 mm) steel pipe is fitted through the curved pipe so that it projects along the central axis. This pipe will be the means of injecting compressed air into the well and should have a strong, water- and air-tight, welded joint with the curved pipe. The bottom should project about 300 mm below the base of the swan-neck and have a threaded end capable of taking the remaining air injection pipes. The top end of the pipe should have a compressed air coupling. The swan-neck is fitted to the top of a length of pipe that will form the rising main. These should normally be steel, but if the pipe below the swan-neck is shrouded by the well casing, then suitable plastic pipe may be used. The air injection pipe projects down the centre of the rising main to about 0.5–1 m above the bottom of the rising main. It is important that a gap of at least 2 m is left above the bottom of the borehole. Normally the gap is much greater than this minimum. The bottom 2 or 3 m of the air injection pipe should be fashioned into a slow taper and be perforated by a large number of small holes. The efficiency of an airlift pump depends on the depth of submergence below the water level. Under ideal conditions, two-thirds of the injector pipe should be submerged during pumping.

Use the information in Table 6.3 to calculate the length of the rising main you need, allowing for the fact that the drawdown when pumping will reduce the submerged depth. The air should be injected through a length of perforated pipe and not simply an open-ended pipe. The large number

Table 6.3 Optimum depth of submergence for airlift pumping.

Total lift (m)	Optimum submergence (%)
10	47–67
20	46–65
30	45–63
40	44–62
50	43–60
60	43–58
70	43–56
80	42–55
90	41–54
100	40–52
120	38–50
150	36–46
200	34–40

of holes ensures that the air and water mix properly to produce a smooth pumping flow. Air injected through the open end of a plain pipe forms a series of large bubbles that delivers the water at the surface in surges and is much less efficient. Airlift pumping saturates the pumped water with oxygen that may change the groundwater chemistry, thereby limiting the value of any samples taken during the pumping test.

Safety considerations cannot be stressed enough! When you use an air compressor to pump water, ensure that all the air delivery connections use the correct types of connector and hose, and that they are the right size and rating for the air flow and pressures being used. It is very important that all the pipework used to inject the air into the borehole is capable of withstanding these pressures, otherwise they may explode. It is safest to use suitable steel pipes; other materials such as plastic or copper can only be used if they are entirely enclosed within the borehole casing and cannot endanger anyone should an explosion occur. It is important to make sure that all the air pipes are tied down to pegs or stakes, as well as any flexible hoses used to carry the pumped water/air mix from the wellhead.

6.4 On-Site Measurements

The two most important measurements made during a pumping test are the pumping rate (or discharge) and the changes in water levels in both the pumping well and the observation wells. Other measurements include monitoring spring flows, rainfall measurements to detect recharge, and atmospheric pressure to identify any barometric effects in the groundwater-level record.

6.4.1 Measuring the discharge rate

A weir tank, as shown in Figure 6.3, is the best method for measuring the discharge rate, as it allows you to see the water and identify anything that may affect the flow readings. For example, a new borehole may produce sand that will wear the impellors in a water meter, thus making the readings too low. In a weir tank, however, the sand will settle out and will not affect its operation, provided that the sand is removed regularly.

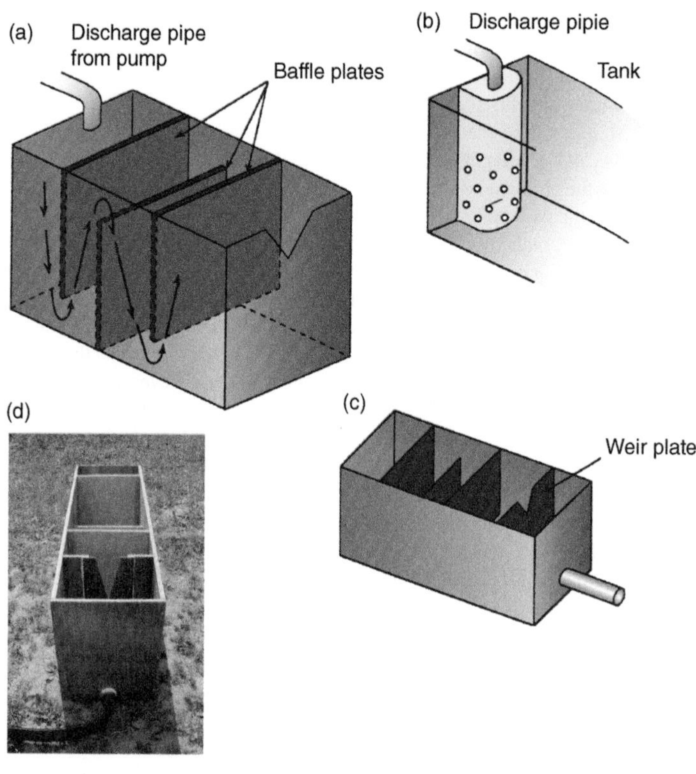

Figure 6.3 *Discharge from a pumping test can be measured using a weir tank such as those shown here (for explanation of parts (a) to (c) see text). The wooden tank in photograph (d) has been made to measure small flows. It has five sections; the first four are separated by baffles that either extend from the base or are suspended above the base of the tank. The fourth and fifth sections are separated by a ¼ 90° weir plate, over which the water flows to discharge into the fifth section, where it is carried to a discharge point through a hose.*

A weir tank uses the same principles to measure flow as a thin-plate weir (see Section 5.4.2) and may have either a v-notch or rectangular weir. The notch may be cut into one end of the tank so that water can flow over it out of the tank, as in Figure 6.3a. The water is pumped into the end opposite the notch and then flows through some baffles that will make sure that the flow is laminar by the time the water gets to the v-notch. You can use a wooden board or perforated metal plates to make the baffles. Even a 25-L drum with holes knocked in it will work if the discharge pipe goes into it (Figure 6.3b).Alternatively, the weir plate is fixed inside the tank with the water being discharged through pipes, which gives more control over where the water goes (Figure 6.3c). A weir tank operates in exactly the same way as a weir plate in a stream. In this case, the water level is measured closer to the weir plate, but accuracy is not lost as the tank is relatively wide compared to the stream installation. The head above the weir is measured at a point behind the weir equivalent to four or five times the depth of water flowing over it. Use Table 5.1 to calculate the discharge rate from the height of water in the tank. The accuracy varies according to the tank size and the discharge rate. A typical tank some 4 m long by 2 m wide and 1.5 m deep will provide accurate measurements up to some

1750 m^3 d^{-1} (c. 20 L s^{-1}). Larger discharges may use more than one tank connected in parallel with pipes at least 200 mm in diameter. Smaller tanks may not measure flows as accurately as large ones, with a decrease in accuracy of up to 5% being likely. Similar reductions in the accuracy are caused by the tank filling with sediment.

A water meter should not be used with airlift pumping as it will record both the water and air flows and give a false high reading. Use a weir tank with the air/water mixture first passing through another tank to allow the air to come out of the mixture before it is measured by the weir tank.

Take time to set up the weir tank correctly at the start of the test. Use a spirit level to ensure that it is absolutely horizontal, otherwise the discharge readings will be wrong. When the tank is full of water it may be heavy enough to sink into soft ground; for example, a tank containing say 5 m^3 of water weighs more than 5 tonnes. Do not risk the tank moving during the test to spoil your readings! Build a temporary platform using large timbers such as railway sleepers to hold the tank on sites with soft ground. Finally, fill the tank with water to the bottom of the weir notch so that it will measure the discharge from the start of the test. Even with this precaution, however, it will take a short time for the water in the weir tank to build up to a constant level so that the full discharge is measured.

Make a record of the discharge rate by installing a pressure transducer in the weir tank. If this is not possible make a note of the water level in the weir tank every 10 minutes for the first hour, then every 30 minutes for the next two hours and several times a day throughout the test. If you are using a water meter you will also have to take the readings at a similar frequency.

Use pipes or flexible hoses to run the water at least 100 m away from the pumping well into a suitable ditch or stream to minimise recirculation. Use a greater distance if you are pumping at high rates. If the water soaks into the ground and flows back to the well it will give a falsely optimistic yield and produce unrealistically high values for the aquifer hydraulic conductivity and storativity. If you have to use a ditch, make a temporary lining out of polythene sheeting. Figure 6.4 shows an example of a weir tank setup in a pumping test. The water is pumped by an electrical submersible pump hung on a rising main and suspended on a pipe clamp that is visible near the hydrogeologist's feet. The rising main continues vertically some 1.75 m, then goes through a 90° bend to run horizontally for some 2 m to a wheel-valve and then a further 2 m to a second 90° bend, where the pipe falls vertically into the end of a weir tank. The wheel-valve was used to control the pump so that the required discharge rates were achieved. The weir tank is about 4 m long and is divided into five sections by three baffles and the weir plate. The first and third baffles stretch from the top of the tank to about 300 mm above the bottom, with the second one sticking up from the bottom. As it flows under and over the baffles the water flow becomes laminar, before being measured by a 90° v-notch. It then discharges into the last chamber and flows out through a short pipe into a temporary ditch lined with polythene sheeting. When the picture was taken, the hydrogeologist was measuring the pumping-water level using a dipper.

Make sure that any water meter lies in a straight piece of pipe for a distance each side of the meter equivalent to at least ten times the pipe diameter, with a minimum length of 1 m. Arrange the pipes you use to ensure that the meter will remain full of water during the test; otherwise the readings will be too high. Check its accuracy using the 'jug and stopwatch' method described in Section 5.3.

6.4.2 Water level measurements

Water level measurements are taken at the same frequency for all types of pumping test. Some readings will be made using a dipper, so make sure that a dip-tube has been installed. When the pump is first switched on, the water level in the well or borehole will change rapidly. Measurements are made very frequently at first and then gradually tail off as the test proceeds. Table 6.4 gives a summary of the frequency for water level measurements and Figure 6.5 shows an example of the type of field sheet used to record test data. Modify the field sheets to suit your particular needs. It is easiest to have one fastened to a clipboard with spare copies kept clean and dry in a polythene bag.

Figure 6.4 *Arrangement for controlling and measuring pump discharge during a pumping test.*

Very few if any pumping tests are carried out with all the readings taken at exactly the correct time. If you miss one, record the *actual* time you took the measurement. Do not forget that you have to start from the beginning with frequent readings each time you change the pumping rate, such as in a step test. When a test goes on for several days, water levels are measured manually only three or four times a day. When the pump has been turned off the recovery is measured at the same frequencies as at the start of pumping.

This table is based on usual practice. Exact frequency is not critical after the first hour provided that data provide an even spread when plotted with a logarithmic time scale. Note: this frequency of readings should also be used at the start of every stage of a step test and for recovery tests when the pump has been shut down.

Table 6.4 Frequency of water level readings during pumping tests.

Time since start/shutdown of pumping	Frequency of readings
0–10 minutes	1 minute
10–60 minutes	5 minute
60–100 minutes	10 minutes
2–5 hours	30 minutes
5–12 hours	1 hour
12–36 hours	2 hours
36–96 hours	6 hours
96–168 hours	8 hours
168 hours to the end of the test	12 hours

PUMPING TEST FIELD SHEET							
Pumping test location				Description of datum point			
Observation point							
NGR				Elevation of datum point			
Date	Time	Elapsed time		Depth to water (m)	Drawdown (m)	Meter reading	Comments
		Minutes	Hours				
		0					
		1					
		2					
		3					
		4					
		5					
		6					
		7					
		8					
		9					
		10					
		15					
		20					
		25					
		30					
		35					
		40					
		45					
		50					
		55					
		60	1				
		70					
		80					
		90					
		100					
		120	2				
		150					
		180	3				
		210					
		240	4				
		300	5				

Figure 6.5 *These field sheets are used to record the water level and discharge measurements taken during pumping tests. NGR stands for National Grid Reference, the UK system of map referencing.*

Try to ensure that the water level readings are taken at the same instant on the pumping well and all the observation wells. Synchronising watches before the start of the test may enable this to be achieved, or have one person managing the time and signalling to the others with a whistle, for example.

Water level readings in the pumping well are taken using a dipper or recorded using a data logger system (see Sections 4.1 and 4.2) which should be installed in a separate dip tube. It needs practice to use a dipper to measure rapidly changing water levels. Take advantage of the pre-test pumping to try it out, so that you start your first proper test with confidence! Ideally, install a data logger system in the pumping well and all the observation boreholes so that you only have to make a few check readings with your dipper. In theory, the frequent readings at the start of a test could be taken by pre-setting the data logger. However, this forces you to start the test at a pre-set time. Delays of a few minutes are common and this can mess up the whole set of readings! The best compromise is to set the data logger to record every 10 minutes for the duration of the test and use a dipper to take the more frequent readings over the first two hours of the test. It is easy to merge the two data sets once the measurements have been input into your computer.

Water levels in the observation wells are recorded in the same way and at the same frequencies in order to provide the data for analysis. In practice, the first few readings in observation boreholes rarely show any response to the pumping except where the borehole lies less than 50 m from the pumping well, so do not worry if your first few readings are all the same. This does not mean, however, that you need not bother to take these readings! Plot a graph of water levels against time on a semi-log scale as soon as the slow-down in measurements permits; you can use your laptop or tablet to do this if you like. It will be easier to find the cause of any anomalies at this stage, rather than after the test has been completed. Section 4.9 gives some of the external factors that can influence your readings.

6.4.3 Sampling during pumping tests

To take water samples during a pumping test, have a sample tap fitted at the top of the rising main. Sample a new borehole near the start of pumping and at the end to see if there have been any changes in quality over the test period. If only one sample is needed, take it at the end when the borehole has been thoroughly purged.

6.5 Pre-Test Monitoring

It is usual to take background readings of groundwater levels, spring flows, rainfall, and barometric pressure starting a few days and occasionally weeks before the start of the test, to provide a baseline so that the impacts caused by the pumping can be recognised. The minimum period is three days and longer for high pumping rates. The size of the area for this monitoring is related to the pumping rate. Table 6.5 shows values for different pumping rates as well as the recommended duration of the test period. Some tests may require new observation boreholes to be drilled to provide a good quality of data or to monitor specific features during the test such as wetland areas. Observation boreholes drilled to obtain data for determining the aquifer hydraulic properties should normally lie between 25 m and 300 m from the pumping borehole, and never closer than 5 m.

6.6 Test Set-up

Make sure that everything is ready at least one day before the start of the pumping test. Once the pump has been lowered into the borehole, connect a valve on the rising main and a pipe to carry the water away. Fit a dip-tube and then try out your dipper to make sure it will fit. If you are intending to use a data logger you should install it now, ideally down a second dip-tube and ensure that it is

Table 6.5 Size of survey areas and duration of pumping tests.

Discharge rate ($m^3 d^{-1}$)	Radius of survey area (m)	Duration of pumping (h)
>20	100	8
20–100	250	8
100–500	500	24
500–1000	1000	48
1000–2500	1500	96
2500–5000	2000	168
>5000	2000–4000	240

working satisfactorily. Take a water level reading and record it in your notebook. This rest-water level will enable you to decide whether full recovery has taken place by the time the pumping test starts. Install the weir tank or flow meter. Make sure that the tank is horizontal and fill it with water when you test the pump before the step test starts. Set up the discharge arrangements, making sure that re-circulation is not possible.

You are now ready to assess the maximum pumping rate and calibrate the control valve for a step test. Before turning on the pump, fully open the valve and then close it, counting the number of turns of the wheel between these two positions. Write this value in your notebook. The relationship between the discharge rate and the number of turns to open the valve is not linear and you are likely to find that all of the control lies within the last turn.

Read the water meter if you are using one. Next, open the valve fully and turn on the pump. Check the water levels in the borehole and continue pumping until the fall in water levels has slowed down. By this time the weir tank should be full and the discharge rate should have stabilised. It is difficult to give precise guidance about how long this will take. With freely yielding wells it may be only 10 minutes, while with others it may be well over half an hour. Record the discharge rate in your notebook from the weir tank or using the water meter and a stopwatch. This will be the maximum rate of the step test and the rate for the constant rate test. Use paint or an indelible pen to mark both the valve wheel and an adjacent fixed object (e.g. the pipe) to act as a reference point for controlling the other valve settings that are needed to achieve each pumping rate. Make a sketch in your notebook or take a photograph on your mobile phone to remind yourself how it looks.

Once you know the maximum pumping rate, divide it by the number of steps to define the other pumping rates as equal fractions of the maximum rate. Next, gradually close the valve so that the flow rate reduces to the next highest pumping rate. If you are using a weir tank, work out in advance what height the water surface should be at the measuring point for each of the pumping rates and ask a helper to monitor the water level as you close the valve. Do not hurry unduly! Remember that there is a time lag between you closing the valve and the reduced flow rate being measured in the weir tank. When you have achieved this second pumping rate, make a note of the relationship between the mark on the valve wheel and the mark on the fixed point, again making a sketch in your notebook and/or taking a photograph. Remember that it may take much less than half a turn of the valve to achieve the flow reduction you want. The process is then repeated for each successive flow rate, always making careful notes on the position of the valve wheel. Once you have achieved the last and smallest rate, leave the valve setting, as this will be the pumping rate for the first step.

6.7 Step Tests

Data from a step test defines the relationship between pumping rates and drawdown. This is termed the *specific capacity* and is measured in terms of the pumping rate/unit drawdown, for example in cubic metres per day per metre. Drawdown is made up of two elements: the first results from the resistance to flow through the well face and is termed *well losses*, and the second is controlled by the aquifer hydraulic conductivity and is called *aquifer losses*. Well losses increase with the higher abstraction rates and form a greater proportion of the total drawdown because the flow into the well becomes steadily more turbulent. Eventually, a point is reached when the water level falls with no further increase in the pumping rate, defining the maximum yield of the well. Specific capacity can be thought of as a graph of pumping rate plotted against drawdown, as shown in Figure 6.6.

A step test involves pumping a well from rest for several equal periods at rates that increase by more or less equal increments. Four steps are usual and three is the minimum number needed to give useful results. Each pumping step needs to be long enough for the water level to stabilise before the pumping rate is increased. Each step is commonly 100 minutes, resulting from a method of analysing the data involving plotting the drawdown on semi-logarithmic graph paper. Drawdown is calculated by subtracting the stable pumping-water level achieved at the end of each pumping rate from the rest-water level measured before you started pumping, as shown in Figure 6.6a. A specific capacity curve (Figure 6.6b) is constructed by plotting the pumping rate against drawdown. At higher rates of pumping, the amount of drawdown increases, indicating that the maximum yield of the well is being approached.

The procedure is simple. Start the pump for the first step with the valve set in the position it was left at the end of the calibration pumping on the previous day. At the same instant, start your stopwatch and commence taking water level readings at the frequency given in Table 6.4. Also measure the discharge rate, as described in Section 6.4.1. At the end of the period you have chosen for that step, quickly open the wheel-valve to the setting that will provide the pumping rate for the second step. Re-start your stopwatch at the same time and begin taking water level readings at 1-minute intervals, as set out in Table 6.4. The remaining steps are managed in the same way. At the end of the final step, switch off the pump and carry out a recovery test (see Section 6.9). Alternatively, you can install a data logger set to take a water level reading every minute, allowing you to concentrate on controlling the flow rates. Minute readings will not provide too many to easily manage the data set on your laptop or pc.

6.8 Constant Rate Tests

A constant rate test should be at the maximum sustainable pumping rate obtained during either the pre-pumping calibration test or the step test. Ideally, it should not be less than the anticipated operational hourly pumping rate. This usually means that the test will be at a higher daily rate than the operational pumping, thereby providing rigorous testing of both the borehole and aquifer so that the long-term sustainable yield and impacts can be assessed with greater confidence. The duration of the constant rate pumping test depends on the rate of abstraction, as shown in Table 6.5.

Before the pump is started, make sure that you have taken a water level reading in the test well and all the observation boreholes, besides readings of any springs you may be monitoring. Start up the test in the same way as with the step test, making your measurements at the frequencies set out in Table 6.4, and measure the discharge, as described in Section 6.4.1. The exact frequency is not critical after the first hour provided that the data provide an even spread when plotted with a logarithmic time scale. Note that this frequency of readings should also be used at the start of every stage of a step test and for recovery tests when the pump has been shut down.

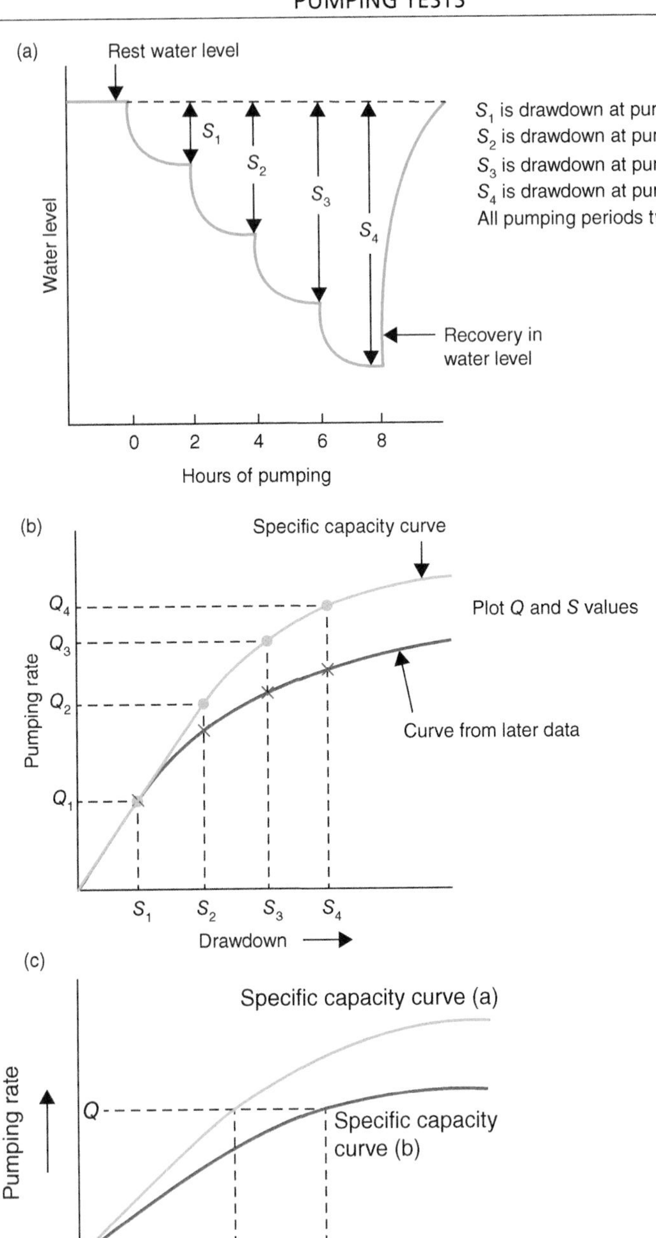

Figure 6.6 *Data from a step-drawdown test have been used to plot a specific capacity curve that defines borehole performance. Diagram (a) shows the water level decline with each step. In (b) two specific capacity curves have been plotted, with the second one produced from a second test carried out at a later date. It can be seen that the well performance has decreased with a greater drawdown for the same pumping rate as shown in (c).*

Figure 6.7 shows the considerations you need to make when planning a pumping test to assess the environmental impact of the abstraction. Following construction of a new borehole at B, a pumping test is required to ascertain the effects on local sources. A survey reveals spring-lines S, S_1, and S_2, a shallow well at A and a borehole at C. The springs from the perched water table would not be at risk, but careful monitoring may still be needed to demonstrate that any reduced flows are due to limitations of the aquifer and not to test pumping. If borehole B receives part of its yield from the upper aquifer, then well A and spring-line S_1 are at risk. Pumping from the lower aquifer could affect spring-line S_2 and borehole C, besides the baseflow of the stream. The extent of these effects would depend on the abstraction rate, the aquifer hydraulic properties and the available groundwater resources.

6.9 Recovery Tests

A recovery test follows on immediately after a step test or constant rate test and usually continues until stable water levels have been achieved. The only measurements required are the changes in water levels in the test well and the observation boreholes at the frequencies set out in Table 6.4. Close the valve to make sure that water cannot drain back down the borehole from the rising main, as it will affect the rate of rise in water levels. Recovery tests provide information on the aquifer properties and provide a useful check on the pumping test data. It is good practice to measure the recovery in water levels after a pumping test for this reason, and it has the further advantage of being relatively cheap to collect, especially if you use a data logger.

6.10 Pumping Test Analysis

The data you have collected during the pumping test should be subjected to mathematical analysis to calculate the aquifer properties and the specific capacity (or efficiency) of the well. A wide range of methods are described in most of the general textbooks in the References list and in great detail by Kruseman and de Ridder (1994). A number of analytical methods are described here, along with some tips on the hydrogeological interpretation of the data. The data are plotted as graphs to make solving complex equations easy. This can be done either by hand or by using a computer standard

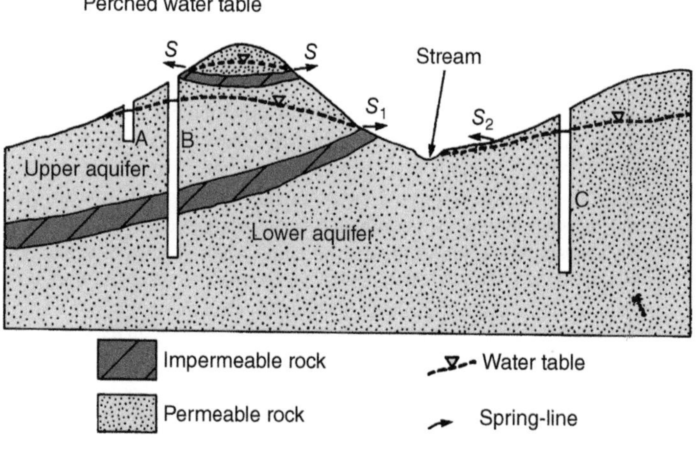

Figure 6.7 *Planning a pumping test to assess the environmental impact of a proposed abstraction requires consideration of local hydrogeological factors such as shown here.*

spreadsheet. Alternatively, there are a number of specialist computer packages for analysing pumping test data, or you can simply use a spreadsheet to plot the graphs and solve the equations yourself. This makes the modern hydrogeologist's life easy, with data downloaded from field instruments and analysed in the computer with no more effort than pushing a few buttons! However, reliance cannot be placed on the results unless there is scrutiny of the data. I am a strong advocate of plotting graphs and thinking about the data before the final calculations are made and conclusions are drawn. For this reason, this section includes a description of some of the graphical methods that can also provide useful insights into the aquifer behaviour.

6.10.1 Data validation

The first step is to examine the water level data and decide whether it has been affected by external influences. The most likely causes of external effects are barometric changes, tidal effects, pumping from other wells and short-term rainfall recharge (see Section 4.9). Data from the period of pre-test monitoring will help you to quantify the extent of these effects and give you the means to correct the field data prior to analysis. You ignore these effects at your peril!

Calculate the barometric efficiency, as described in Section 4.9.2, and use barometric data for the test period to adjust your water level record to a constant atmospheric pressure. Tidal effects are likely to be small. Compare the variations in groundwater levels measured prior to the test with the state of the tide by looking at hourly values taken from tide tables. There is likely to be a constant time lag between high tide and the peaks in water levels. Use this relationship to correct the data throughout the test. Remember that the time of high tide changes for each tide and that spring tides will cause a greater effect on water levels than neap tides, so take account of these variations in calculating your correction factors.

Pumping from other sources can be a problem because of variations in pumping times. If it is not possible for the other abstraction to cease for the period of the test (including the pre- and post-test measurements), try to persuade the abstractor to pump continuously at a constant rate.

The impact of rainfall recharge is assessed using the pre-pumping monitoring record. Also consider the geology, to determine whether local recharge is feasible. Compare any fluctuations with rainfall records for the same period and develop an approximate relationship by plotting a simple graph. Do not forget that rainfall is usually associated with falling barometric pressure, which will cause a rise in water levels that has nothing at all to do with recharge!

6.10.2 Analysis of non-equilibrium data

The Thiem or equilibrium well equation is discussed in Section 3.5.1. The stable drawdown at the end of a constant rate test can be analysed using this equation to calculate a value for the hydraulic conductivity. All the methods described here use graphical methods to solve a complex equation, with drawdown in metres plotted on a normal scale against time in minutes on a logarithmic scale. Computer systems are available for pumping test data analysis, but these have been ignored here so that the relationships can be better seen. A large number of analytical methods are available that use the rate of change in water levels to calculate both the hydraulic conductivity and storativity. The latter is made possible because the volume of water in storage is changing throughout the test, although it requires observation borehole data, not just pumping borehole information.

Every pumping test analysis method makes a number of assumptions about the aquifer. However, in many cases the aquifers differ from these assumptions. It is important to consider how closely the real conditions meet the assumptions so you can select the appropriate method of analysis. For more information see Kruseman and de Ridder (1994).

6.10.3 External influences on pumping tests

A semi-log time–drawdown graph forms a straight line if all the assumptions made about the aquifer are met. The graph deviates from a straight line because of aquifer boundary effects or well effects. By considering the potential of each type of effect, it is possible to make deductions about the aquifer behaviour.

The effects caused by aquifer boundaries can be divided into two types, barrier boundaries and recharge boundaries. Barrier boundaries are where the aquifer effectively ends within the radius of influence of a pumping well, as either the physical limit of the aquifer formation or a fault against low-permeability rocks. Recharge boundaries are where there is a large volume of water available to recharge the aquifer once the cone of depression has extended to reach it. This may be a large surface-water body or an adjacent aquifer with a significantly greater storativity. Figure 6.8 shows different types of aquifer boundary that affect the time–drawdown graph. In Figure 6.8a, data from a pumping test in an aquifer with no boundary effects form a simple straight-line graph, with all the pumped water being derived from aquifer storage. If the pumping well is near a river, a lake, or even the sea, when the developing cone of depression reaches this water body part of the water flowing to the well is obtained from this water body, reducing the rate at which aquifer storage is developed. This causes the rate of decline in pumping water levels to decrease, as shown in Figure 6.8b. The converse situation is where the well is close to the edge of the aquifer or a barrier to groundwater flow, such as a mineralised fault. In this case (Figure 6.8c), the rate of decline in water levels increases, as the water pumped from the aquifer storage is taken from a restricted volume of rock.

During the first part of a test, most of the water is pumped from that stored in the well. This affects the rate of fall in water levels, especially where the inflow rate to the well is small, due to either a low hydraulic conductivity or a clogged well face. Schafer (1978) suggested a method to calculate the significance of well effects on the early data from a pumping test. The Schafer equation is

$$t_c = \frac{0.017(d_\mathrm{c}^2 - d_\mathrm{p}^2)}{Q/s}$$

where

t_c is the time in minutes when the casing storage becomes negligible
d_c is the inside diameter of the casing in millimetres
d_p is the outside diameter of the rising main in millimetres
Q/s is the specific capacity of the well in cubic metres per day per metre of drawdown at time t_c.

Figure 6.9 gives an example of a pumping test on a well in a sandstone aquifer. The effects of well storage can be assessed using the Schafer equation. The internal diameter of the borehole (d_c) is 200 mm, the diameter of the rising main (d_p) is 50 mm and the specific capacity (Q/s) is 89.174 $m^3\,d^{-1}\,m^{-1}$. Substituting these values into Schafer's equation to calculate the duration of the period when the well storage effects are significant, we obtain

$$t_c = \frac{0.017(200^2 - 50^2)}{89.174} = 7.15 \text{ minutes}$$

On the basis of the Schafer analysis it can be concluded that well storage effects are significant during the early part of the test data for a period of less than 10 minutes from the start of pumping. In Figure 6.9 we can see that the data from 10 minutes also plot as a straight line. It represents the flow through the aquifer unaffected by either well effects or boundary effects for about 6 hours. After that, the impact of a barrier boundary causes an increase in the rate of drawdown until the end of the test at 48 hours.

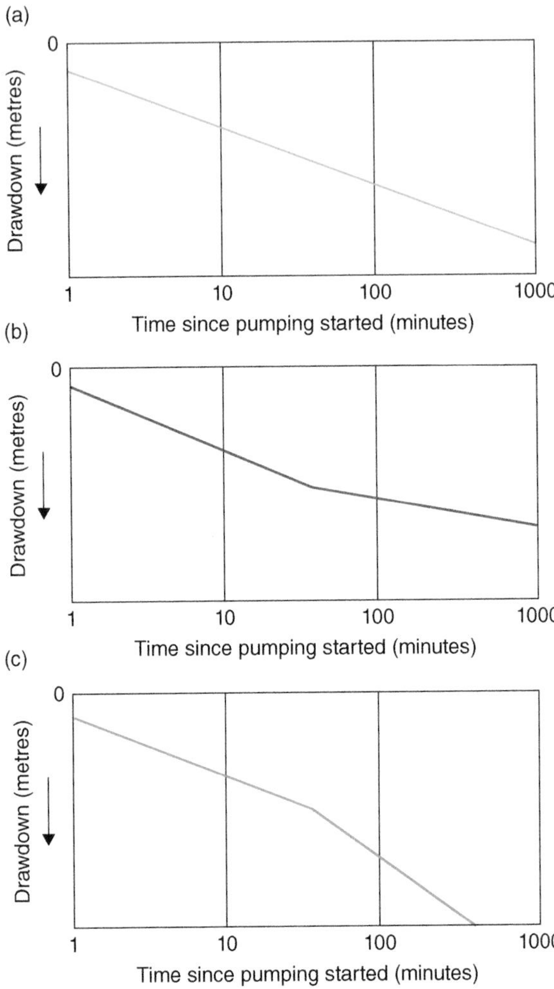

Figure 6.8 *The Cooper–Jacob method of pumping test analysis uses a graph of drawdown on a normal scale against time in minutes on a semi-logarithmic scale and is a useful preliminary way of examining pumping test results.*

Data from a pumping test on a sandstone aquifer have been reviewed in Figure 6.10 to illustrate the effect of barriers on pumping test water levels. The borehole is some 200 mm in diameter and 100 m deep, with casing to 25 m depth in the Upper Coal Measures of northern England. It penetrates mudstones (with a coal seam at 15.5 m) to 30 m depth that overlie mudstone with thin sandstones to 60 m and then sandstone to the full depth of the borehole. All the sandstones are potential aquifers, with the deepest one being dominant. Rocks in the area are faulted and there are abandoned deep coal workings. The data have been plotted to produce the time–drawdown graph, which can be divided into five segments, as shown in the diagram. A Schafer analysis shows that the well storage effects lasted about three minutes, which is represented by segment 1. segment 2, from 3 to 300 minutes of

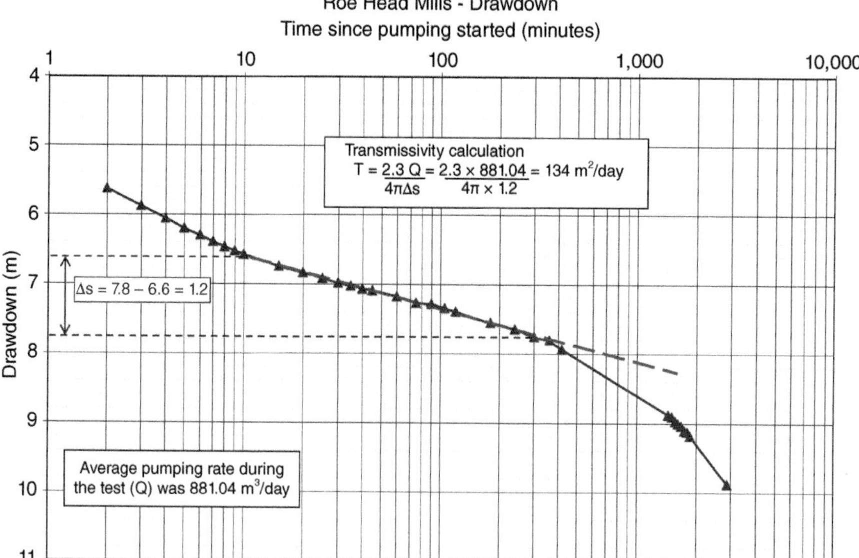

Figure 6.9 *Data from a pumping test at Roe Head Mills at a rate of approximately 880 m^3 d^{-1}, with the effects of a barrier boundary in the later data.*

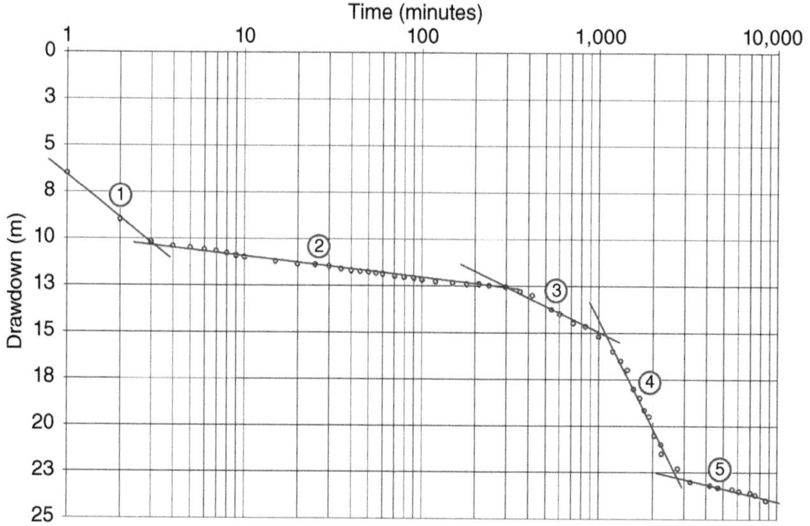

Figure 6.10 *Water level measurements taken during a 7-day pumping test at a rate of 2074 m^3 d^{-1} show several boundary effects.*

the test, represents groundwater flow through the sandstone aquifer and can be used to calculate aquifer transmissivity. The third segment (300–1000 minutes) shows an increased rate of drawdown caused by the cone of depression reaching a barrier boundary that is assumed to be one of the nearby faults, as data from several observation boreholes would be needed to compute the direction and distance of this boundary. A second barrier boundary (probably another fault) is represented by the fourth segment (1000–2400 minutes). During the final part of the test, the rate of drawdown decreased significantly, suggesting that a recharge barrier had been intercepted, or possibly the barrier effects have finished with the water level at about 25 m depth.

6.10.4 Step test analysis

The simplest way of analysing step test data is to plot a graph of the specific capacity, as shown in Figure 6.6. More sophisticated methods enable the different factors that affect drawdown to be studied in more detail. Cooper and Jacob (1946) suggested that the drawdown in a pumping well can be expressed by the following equation:

$$s = BQ + CQ^2$$

where
s is the drawdown in metres
Q is the pumping rate in cubic metres per day
B is a co-efficient related to laminar flow
C is a co-efficient related to turbulent flow.

Bierschenk (1964) proposed a graphical method to solve Jacob's equation based on the relationship where s_w is the drawdown in the pumping well in metres, Q is the pumping rate in cubic metres per day, BQ is the formation loss (i.e. the amount of drawdown resulting from the aquifer permeability in metres) and CQ^2 is the well loss (i.e. the drawdown resulting from the hydraulic resistance of the well screen/well face in metres).

$$s_w = BQ + CQ^2$$

If Bierschenk's equation is written in the form $\frac{s_w}{Q} = B + CQ$ it can be solved graphically by plotting s_w/Q against Q to produce a straight line, where B is the intersect on the y-axis and C the slope of the line.

Data from a step test carried out on a borehole in a gravel aquifer in Paramali, Cyprus have been plotted in Figure 6.11 to illustrate this procedure. The first graph (Figure 6.11a) shows the drawdown data from the test. When the data are plotted in the second graph (Figure 6.11b), the points do not easily fall on a straight line. This is very common and results from a different stage of equilibrium being reached in each step. The values for the constants in the Bierschenk equation are taken straight from the graph, with $B = 1.40 \times 10^{-3}$ dm^{-2} and $C = 1.23 \times 10^{-6}$ d^{2}m^{-5}. It is possible to calculate the efficiency of the borehole at each pumping step using the following equation:

$$\text{efficiency} = \frac{BQ}{BQ + CQ^2}$$

The results summarised in Table 6.6 show that the efficiency decreases with increasing pumping rates, which is caused by turbulent losses around the well screen and frictional losses inside the casing, rather than the simple inefficiency of the well design. In addition, it is possible to use Logan's formula to calculate the transmissivity (T), using the data from Table 6.6, which in this case is 539 m^2 d^{-1}.

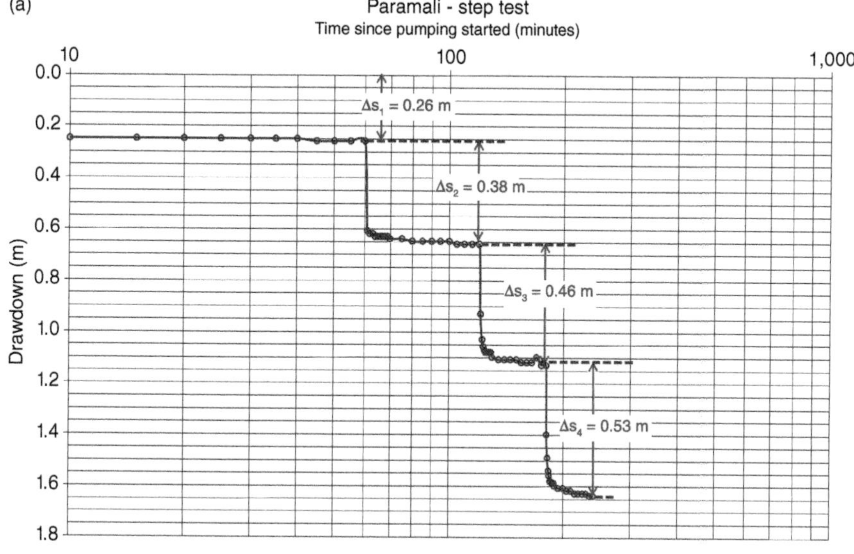

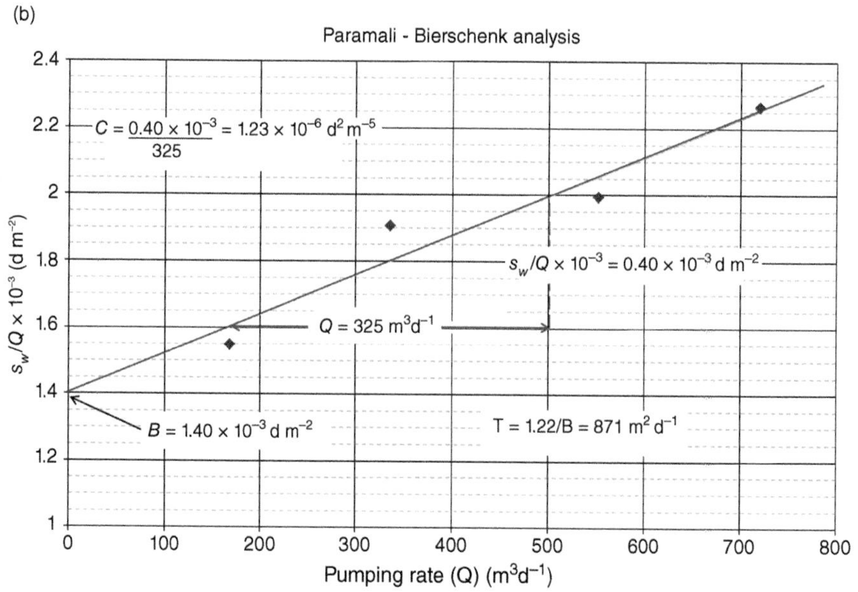

Figure 6.11 *A graphical method to solve Bierschenk's equation.*

6.10.5 Constant rate test analysis

A simplified method of analysing constant rate tests was proposed by Cooper and Jacob (1946) using a semi-log plot of the time–drawdown data. The method was developed for confined aquifers only, but it can be applied to unconfined conditions provided that the drawdown is small compared to the saturated thickness of the aquifer. The Cooper–Jacob equation is

5. Rainfall, Springs and Streams
6. Pumping Tests
7. Groundwater Chemistry
8. Recharge Estimation

Table 6.6 *Summary of results from the test at Paramali, Cyprus.*

Step	Δs_w (m)	s_w (m)	Q ($m^3\ d^{-1}$)	Efficiency (%)
1	0.26	0.26	168	78.4
2	0.38	0.64	336	64.4
3	0.46	1.10	552	52.4
4	0.53	1.63	720	45.8

$$T = \frac{2.30Q}{4\pi\,\Delta s}$$

where

T is transmissivity in square metres per day
Q is the discharge rate in cubic metres per day
Δs is the slope of the straight line time–drawdown graph over one log cycle on the time axis.

In these semi-logarithmic graphical methods, the slope of the straight line is used in the equation and is taken to be the difference in the drawdown over one log cycle.

The Cooper–Jacob method also allows the storativity to be calculated using the following equation:

$$S = \frac{2.25T\ t_0}{r^2}$$

where

S is the storativity (dimensionless)
T is the transmissivity in square metres per day
t_0 is the intercept of the time graph at zero drawdown in days
r is the distance (in metres) of the observation well from the pumping well.

The Cooper–Jacob method was used to analyse data recorded in an observation borehole during a pumping test on a well in a sandstone aquifer. The test data have been plotted on a semi-log scale in Figure 6.12 and approximate to a straight line. It is possible to analyse these data using the Cooper–Jacob equation shown above. In this example, $Q = 1175\ m^3\ d^{-1}$ and $r = 275$ m. Values for parameters taken from the graph are $\Delta s = 0.51$ m and $t_0 = 0.028$ d.

$$T = \frac{2.30Q}{4\pi\,\Delta s}$$

therefore

$$T = \frac{2.30 \times 1175}{4\pi \times 0.51} = 421.7\ m^2 d^{-1}$$

$$S = \frac{2.25T\ t_0}{r^2}$$

therefore

$$S = \frac{2.25 \times 421.7 \times 0.0278}{275^2} = 3.49 \times 10^{-4}$$

6.10.6 Recovery test analysis

Recovery data can be analysed in a similar way to pumping data. In this case, residual drawdown (i.e. the difference between the well water level and the original static level before the start of the

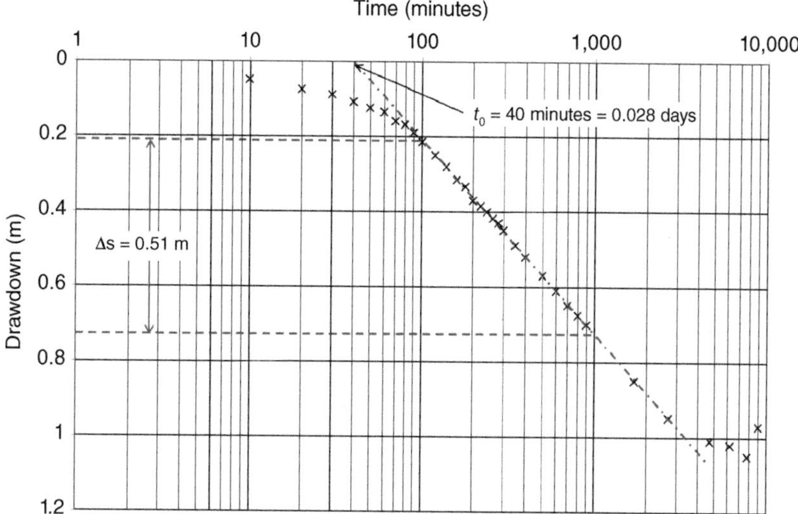

Figure 6.12 *Data from a pumping test are plotted on a semi-log scale to solve the Cooper–Jacob equation.*

pumping period) is plotted against a ratio (t/t') of the time since the pumping test started (t) to the time since the recovery test began (t'), both measured in minutes.

The recovery equation proposed by Thies (1935) is

$$T = \frac{2.30Q}{4\pi\,\Delta s'}$$

where

T is transmissivity in square metres per day
Q is the average discharge rate during the pumping period in cubic metres per day
$\Delta s'$ is the slope of the straight line expressed as metres per log cycle of t/t'.

Figure 6.13 shows an example of a recovery test to illustrate how the method is used. The measurements of recovery once pumping has ceased are called residual drawdown, which is the water level reading below the original rest-water level. The residual drawdown (s') is plotted against t/t'. The graph shows the recovery data from the same pumping test as in Figure 6.12.

In this example, $Q = 1175\ \text{m}^3\ \text{d}^{-1}$ and from the graph $\Delta s' = 0.48$ m

$$T = \frac{2.30Q}{4\pi\,\Delta s'}$$

$$T = \frac{2.30 \times 1175}{4\pi \times 0.48} = 448.0\ \text{m}^2\text{d}^{-1}$$

The analyses in Figures 6.12 and 6.13 give slightly different results for transmissivity and indicate an actual value of a little over 400 $\text{m}^2\ \text{d}^{-1}$ for the aquifer. These slight variations are common and result from external factors and subjective decisions over the best-fit straight line.

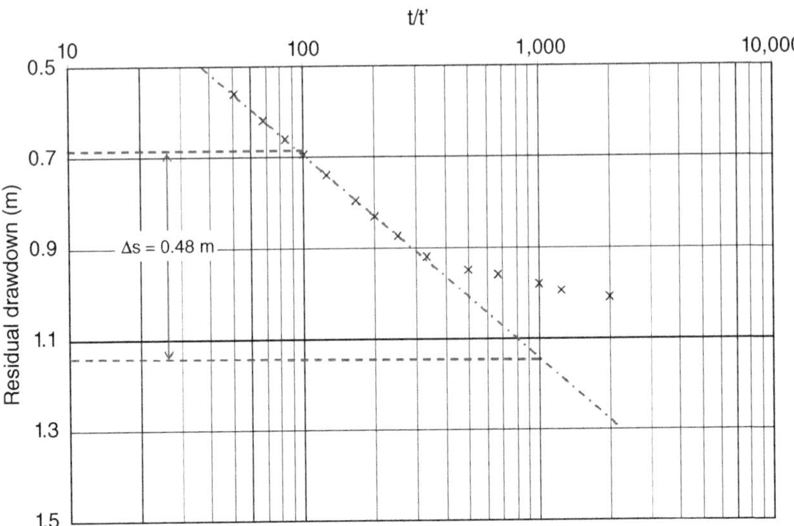

Figure 6.13 *Recovery data from the same pumping test as in Figure 6.12.*

6.11 Tests on Single Boreholes

The tests that come under this heading are usually called slug tests, and strictly speaking are not pumping tests at all! They involve artificially changing the water level in the test borehole and then monitoring the water level as it resumes its original level. They are usually carried out on narrow boreholes with diameters of 50 mm or less.

There are two types of slug test: falling head tests, where the water level is artificially raised, and rising head tests, where water level is rapidly lowered. The change in water level may be achieved by quickly adding or extracting water from the borehole. More commonly, a solid cylinder is either inserted into or removed from the water column to achieve this effect. Slug tests are described by Preene et al. (2000) and Cashman and Preene (2013) with the standard methods for analysing the data.

Slug tests are relatively quick, cheap and more easily carried out than pumping tests and only involve simple equipment. These are their only attractions! The installation of small-diameter boreholes (see Section 9.1) often causes compaction of unconsolidated materials close to the borehole well face which will reduce the permeability. Alternatively, there may be localised fracturing of solid formations close to the piezometer which will locally enhance the permeability. The small volumes of water used for these tests mean that the effect extends only a small distance from the borehole, possibly less than 10 cm, so the measurements do not represent the aquifer as a whole and may only reflect the changes caused by construction. This contrasts with pumping tests, where the rock mass being tested extends beyond the observation boreholes. It may be tempting to use slug tests in hydrogeology, although they should be used sparingly and with great caution. I avoid using them myself because of doubts about the accuracy, discussed above, and prefer to estimate the permeability from the rock type (see Chapter 3) when it is not possible to carry out a pumping test.

6.11.1 Falling head tests

To carry out a falling head test by injecting a known volume of water into a borehole you will need an injection pipe, hose, funnel, bucket, dipper, and/or pressure transducer, stopwatch, and notebook.

Dirty water with material in suspension will clog the test section and affect the test results. Use clean mains water, which should be transported to site in a water bowser (a trailer-mounted tank that holds 0.5–1.0 m^3 of water). The pipe is lowered into the well so that the end is below the water level. The water is injected by opening the valve on the tank and allowing the water to flow into the borehole under gravity or by tipping the water into the funnel attached to the top of the pipe. Do not increase the water level in the well by more than a metre.

The displacement method using a cylinder avoids problems caused by introducing dirty water. You will need the cylinder and a rope to lower it down the hole. Simply lower the cylinder quickly below the water surface to create the rise in water level. Use a cylinder about 75% of the diameter of the test borehole, which will mean that the water level will rise by an amount equal to slightly over half the length of the cylinder.

Water levels should be measured at the same frequency as for a pumping test (see Table 6.5). Start your measurements of the water level at the instant you stop injecting the water or the cylinder is submerged. The easiest method is to use a cylinder with a data logger system set to record the water level at minute intervals. Continue the water level measurements until the pre-test conditions have been re-established.

6.11.2 Rising head tests

The procedure for these tests requires a volume of water to be removed rapidly from the borehole. This is achieved by using either a bailer or a displacement cylinder large enough to remove enough water to lower the level by at least 0.5 m. Using a data logger makes it easy; make sure that you wait until the original water levels are restored. This type of test is likely to produce more reliable results, as the groundwater is flowing into the well and clogging of the well face is far less of an issue.

6.11.3 Methods of analysis

The most commonly used method to analyse the data is based on the work by Hvorslev (1951) using the following equation:

$$K = \frac{A}{FT}$$

where

K is the hydraulic conductivity in metres per second
A is the cross-sectional area of the borehole casing at the water level during the test in square metres
F is the shape factor
T is the basic time lag in minutes (not to be confused with transmissivity!).

Hvorslev recognised that the geometry of the test section will affect the calculation, and introduced the concept of a shape factor. Figure 6.14 gives the shape factors for the most common situations.

The water level below ground (H) is measured at the frequency shown in Table 6.5, and the basic time lag (T) is determined graphically, as shown in Figure 6.15. Calculate values of H/H_0 by dividing the drawdown value (H) by the drawdown at the start of the test (H_0) when $t = 0$. The data plot on a straight line going through the origin (i.e. when $H/H_0 = 1$ and $t = 0$), although sometimes the early points may fall off the line. The procedure involves drawing a straight line through as many of the data points as possible. If it does not pass through the origin, draw a parallel line that does. The basic time lag is the value of t when $H/H_0 = 0.37$. Read this value off the graph and substitute it in the equation. The hydraulic conductivity value will be in metres per second; to convert it to metres per day multiply by 8.640×10^4.

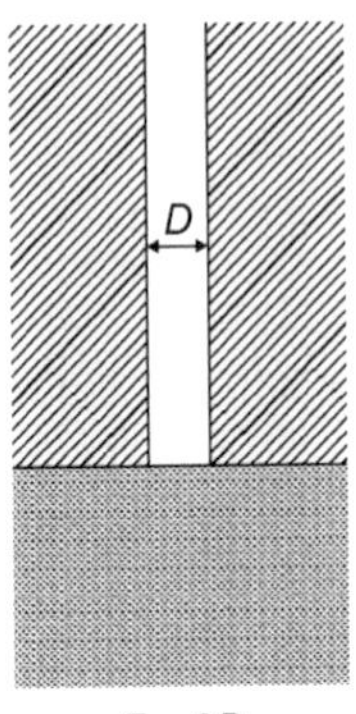

$F = 2D$

Soil flush with bottom of casing at impermeable boundary

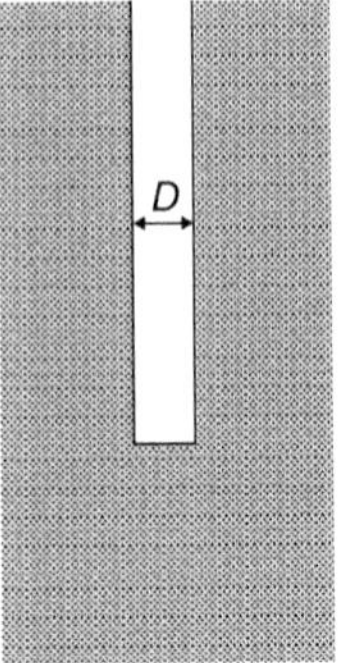

$F = 2.75D$

Soil flush with bottom of casing in uniform soil

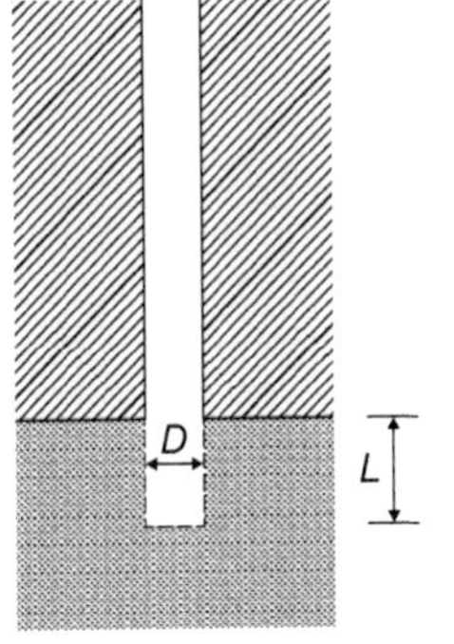

$$F = \frac{2\pi L}{\ln\left[(2L/D) + \sqrt{(1 + (2L/D)^2)}\right]}$$

Open section of borehole extended beyond casing at impermeable boundary

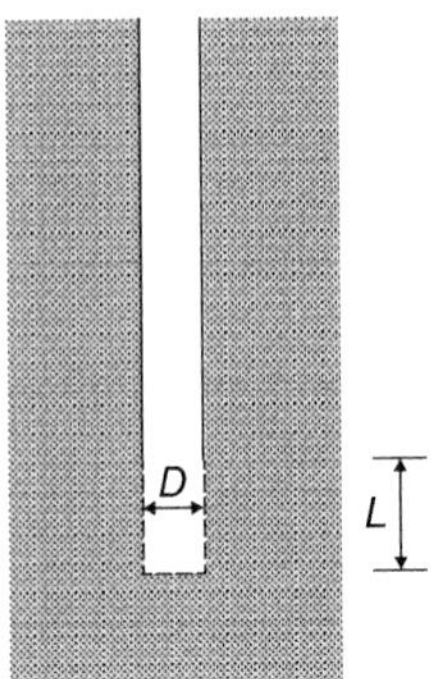

$$F = \frac{2\pi L}{\ln\left[(L/D) + \sqrt{(1 + (L/D)^2)}\right]}$$

Open section of borehole extended beyond casing in uniform soil

Figure 6.14 *Shape factors (F) for calculating hydraulic conductivity from slug tests using the Hvorslev equation (where L is length of test section in metres and D is diameter of the test borehole at the water surface in metres). (Hvorslev 1951/U.S Army/Public domain.)*

In the example shown in Figure 6.15 the borehole is 26.6 m deep and is cased to 14.4 m, giving a test length of 12.2 m with a borehole diameter of 0.1 m. The basic time lag is 11.6 minutes. The shape factor that applies for these tests is defined by the following formula:

$$F = \frac{2\pi \mathrm{L}}{\ln[(\mathrm{L}/\mathrm{D}) + \sqrt{1 + (\mathrm{L}+\mathrm{D})^2}]}$$

where

F is the shape factor
L is the length of the test section in metres
ln is natural logarithm to base e
D is the diameter of the test borehole at the water surface in metres.

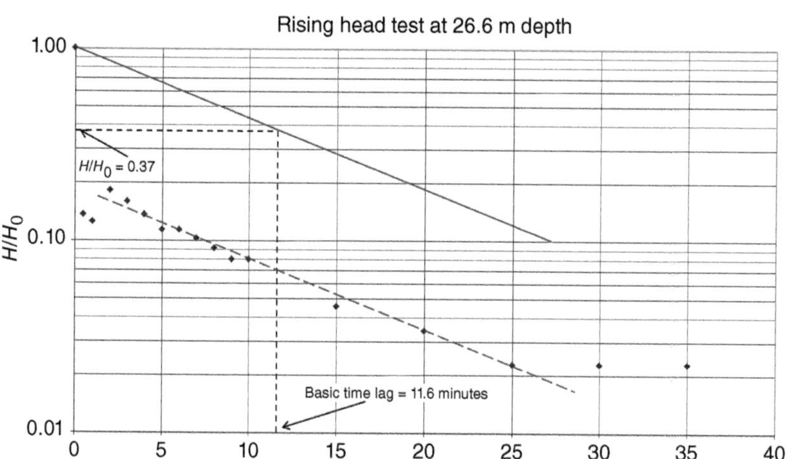

Figure 6.15 *In Hvorslev's method the basic time lag (T) is determined graphically as shown.*

Applying the data:

$$F = \frac{2\pi 12.2}{\ln[(12.2/0.1) + \sqrt{1 + (12.2 + 0.1)^2}]} = 15.64$$

Applying the data to the formula given above:

$$K = \frac{A}{FT} = \frac{7.85 \times 10^{-3}}{15.64 \times 11.6} = 4.327 \times 10^{-5} \text{m s}^{-1} = 3.74 \text{ m d}^{-1}$$

The rocks penetrated by the test section were fractured sandstone and highly fractured mudstone. Compare with the values shown in Figure 3.6 and the value is slightly higher than would have been estimated from the lithologies.

6.12 Packer Tests

Borehole packers are devices that can be inflated to form a seal in a borehole. They are essentially a reinforced rubber hose held on a steel assembly so that the packer can be suspended on a pipe down a borehole. The packer is inflated using compressed air or nitrogen that is delivered down the borehole through small-diameter tubing. Quite often packers are used in pairs so that they are able to seal off a section of borehole, so that tests can be made for the hydraulic conductivity to be calculated or samples to be taken. Packer tests are only carried out in unlined sections of boreholes drilled into solid formations.

A packer test is the term used for hydraulic conductivity measurements using borehole packers. Such tests can be divided into pumping-in tests, where water is added to the borehole, and pumping-out tests, where water is pumped from the test section, rather like a mini-pumping test. Experience has shown that the results from pumping-in tests are frequently affected by the test method (Brassington and Walthall, 1985). The introduction of test water often results in the well face in the test section becoming clogged by particles in suspension. Injection methods often employ high pressures that may fracture the rock, thereby increasing the measured hydraulic conductivity values. Some effects of fracturing are not always obvious when they are limited to microscopic damage. For example, secondary illite growths are common within the pore spaces in some Triassic sandstones and can be broken easily by high water pressures, with the fragments blocking the pore spaces, thereby lowering the hydraulic conductivity value. The data from injection tests are often suspect and I would only ever use pumping-out tests. This also has the great advantage that water samples can be taken easily, allowing a depth profile of the water chemistry to be built up at the same time as the hydraulic conductivity.

Price and Williams (1993) and Walthall (1999) describe the equipment that is used and the tests in some detail. Figure 6.16 shows how the equipment is installed for a pumping test using a double packer assembly. The lower section of pipe is about 50 mm diameter and the rest is 100 mm to

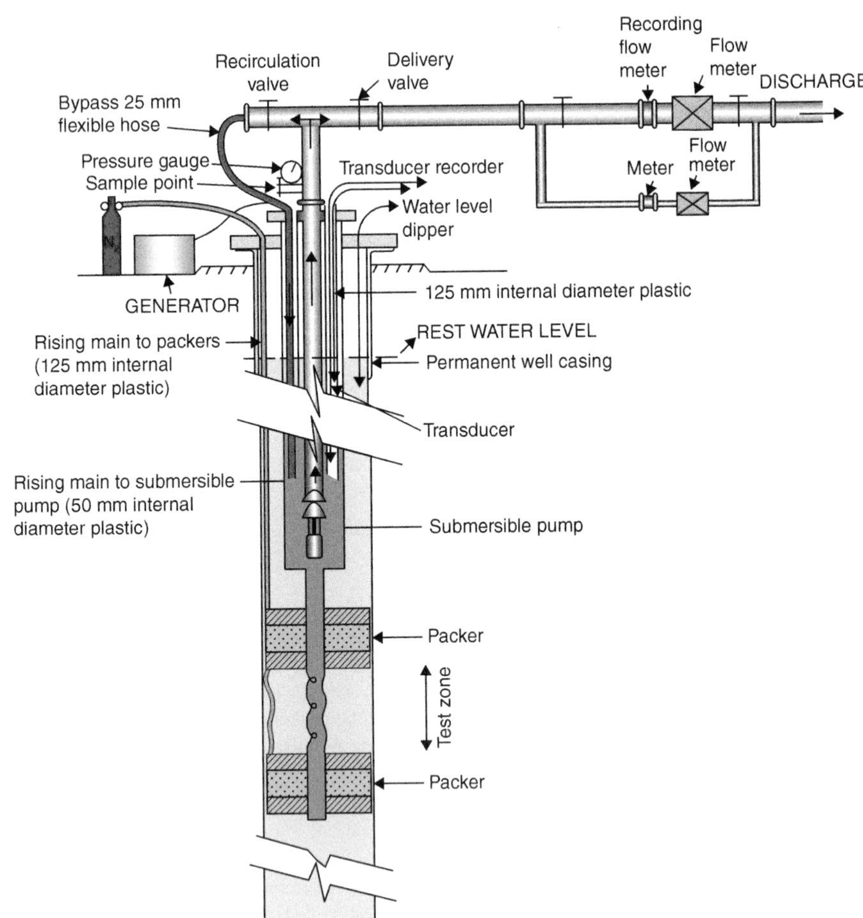

Figure 6.16 *A double packer assembly hung on pipe and installed in a borehole. (Reproduced from Brassington and Walthall, 1985 by permission of the Geological Society.)*

accommodate the submersible pump used for the test. The packers are lowered to a section in the borehole selected from a caliper log or CCTV information, and are inflated using compressed nitrogen or air. Water is pumped from the test section, with the pumping rate controlled by a valve. Low rates of pumping are made possible by an arrangement that recirculates some of the water back into the pumping chamber. This equipment can be used for pumping tests for determining the hydraulic conductivity or for obtaining water samples for analysis. The recirculation system should not be used in sampling. It is important for the packers to seat properly on the borehole wall and provide a good seal. The best way to avoid major fractures and irregularities is to use a caliper log with a CCTV inspection as a guide. The packer assembly is lowered to the selected level, and once it is secured the packers are inflated using the compressed gas supply. Before you turn on the gas, check the water level in the borehole, and repeat the measurement once the packers are inflated; also measure the water level in the borehole above the packers. It is normal for there to be a difference in these water levels, and this is a good sign that the packers have sealed. If you have installed a data logger in the test section measuring levels at frequent intervals (say 30 seconds) you can actually see the levels change as the seal is made. These measurements allow you to construct a depth profile of the heads within the aquifer. This type of equipment has been used for testing boreholes to depths in excess of 400 m.

Pump from the test section, measuring the change in water levels and discharge rates, as described in earlier sections, until steady-state conditions are reached. The data are normally analysed using versions of the Hvorslev equation. However, Walthall (1999) suggests that provided the length of the test section is more than five times the borehole diameter, an average value of the hydraulic conductivity can be obtained from the following equation:

$$K = \frac{Q \ln(1/r)}{2\pi LH}$$

where

K is the hydraulic conductivity in metres per day
Q is the discharge rate in cubic metres per day
r is the radius of the test section in metres
L is the length of the test section in metres
H is the change in head in metres.

When using this equation, make sure that you use natural logarithms (ln) to base e and not common logarithms, which are to base 10.

7
GROUNDWATER CHEMISTRY

Studying the groundwater chemistry will allow you to decide on the suitability of groundwater for drinking or for other uses, or help identify the cause of pollution. An understanding of groundwater chemistry can also tell you a great deal about a groundwater flow system, particularly the relationships between different aquifers, recharge mechanisms, and the relationship between groundwater and surface-water bodies. Information on all aspects of groundwater quality depends on the analysis of water samples taken either by you or as routine source monitoring by organisations such as water companies. An understanding of sampling methods and the equipment used is important in both planning your own sampling programme and evaluating the reliability of the data from other sources.

7.1 Analytical Suites and Determinands

The first step in planning a groundwater quality study is to decide what chemical and bacterial analyses are required. It is usual to carry out these measurements in groups of determinands that together provide a good indication of specific aspects of the groundwater quality. Table 7.1 sets out the determinands in six commonly used analytical suites, from which you can select those that are appropriate to your study. Ideally, a full analysis should be carried out on at least one sample taken from each sample point in your study area, comprising all the determinands in analytical suite 1 and 2, and iron and manganese from analytical suite 3 of the table. This may seem expensive, but it will provide a comprehensive understanding of the groundwater chemistry and whether any pollution is present.

More than 95% of the major ion chemistry is defined by the ions included in analytical suite 2 (Table 7.1) and is of fundamental importance in all investigations (Hem, 1985). Analytical suites 3–6 are often used in pollution studies. If you are unsure about the water being polluted, have a sample scanned with a gas chromatography mass spectrometer (GCMS), which indicates the presence of organics such as fuel oils and pesticides as well as other chemical substances. More detailed analysis may be needed to identify the exact species of any pollutant that may be found.

Take field measurements of those parameters that are likely to change before the sample reaches the laboratory, as well as having them measured in the laboratory. The controlled conditions and equipment needed for many tests means, however, that the majority of measurements can only be made in a laboratory. It is important to ensure that the laboratory you use is able to carry out all the analyses you require using suitable methods and with the level of accuracy needed. Just because they measure a particular ion does not mean that they do it accurately enough to be of use in hydrogeological studies. The laboratory manager will be pleased to provide you with all the information you need to make these judgements. Only use a laboratory that has had its testing procedures accredited by a recognised standards organisation to ensure that the results are accurate to a known standard. Accredited laboratory tests are usually required by environmental regulators.

Occasionally check the accuracy of the laboratory results by including a duplicate of one of the samples, made anonymous with a different name and identification number. If the duplicate results fall within the quoted analytical error, the results for the whole batch of samples are acceptable. Any discrepancies outside the permitted error should be discussed with the laboratory staff with a view

Field Hydrogeology, Fifth Edition. Rick Brassington.

5. Rainfall, Springs and Streams
6. Pumping Tests
7. Groundwater Chemistry
8. Recharge Estimation

Table 7.1 *Analytical suites for groundwater quality assessment.*

Suite	Determinands
Field parameters	Temperature, pH, Eh, dissolved oxygen, electrical conductivity (EC), alkalinity
Major ions	Ca, Mg, Na, K, HCO_3, CO_3, Cl, SO_4, PO_3, NH_3, NO_2, NO_3, total organic carbon (TOC), EC, pH
Minor ions	Al, Cd, Hg, As, I, Mo, Ba, Cr, Ni, Pb, CN, Sr, B, Cu, Br, Li, Zn, Se, Sb, Fe, Mn
Organic	TOC, chemical oxygen demand (COD), biological oxygen demand (BOD), aromatic hydrocarbons, halogenated hydrocarbons, phenols, chlorophenols
Pesticides	Atrazine, simazine, mecoprop, isoproturon, triallate, 2-methyl-4-chlorophenoxyacetic acid, plus others known to be used in the area
Bacteria	Total coliforms, faecal coliforms, colony counts at 22°C and 37°C

Eh is the redox potential or oxidation/reduction potential.
Field measurements are additional to laboratory suites.
Suites 2–6 require laboratory measurements.
The core major ions for an ionic balance of 95% are Ca, Mg, Na, K, HCO_3, CO_3, Cl, SO_4.
An ionic balance should be carried out on suite 2 to check accuracy at 5% or better.
HCO_3, CO_3 may be quoted as alkalinity or $CaCO_3$.
Total nitrogen measures NH_3, NO_2, and NO_3.
TOC is the general indicator of organic pollution.
Fe and Mn should be analysed as both a total and dissolved concentrations.

to repeating some tests and identifying the source of the error. Test your own field procedures using a standard sample that has been prepared in the laboratory. It is taken into the field and tested for the standard field parameters using the instruments you use for your sampling programme. It should be filtered and bottled like a normal sample, before being sent to the laboratory with an anonymous identification number. The analyses should be checked against the normal samples to determine if the sampling method affected the results.

Understanding the way that the laboratory carries out a particular analysis may be important, as it could affect how you interpret the result. There are chemical compounds that are poorly soluble in water and the laboratory may need to dissolve them in strong acids in order to be able to make the analysis. If these chemicals are present in solid form rather than being dissolved in the groundwater they will add to the analytical result and could cause confusion about what is actually in solution in the water. These problems are most likely to exist in pollution studies where compounds such as polycyclic aromatic hydrocarbons (PAH) are almost entirely insoluble and also have a strong tendency for the molecules to be held on the surface of small particles by adsorption. As a result, the analyses for such contaminants may show PAH to be present although none is actually dissolved in the groundwater. The only way to be certain of what is dissolved in the groundwater is for the analysis to be carried out on a pre-filtered water sample. Either carry this out yourself or discuss it with the laboratory manager.

7.2 Sampling Equipment

It is relatively easy to obtain a sample from a supply well from a tap at or near the wellhead. Make sure that the sample is taken from the tap *before* the water has been treated. You will need to use special equipment to take a water sample from a well or borehole on which there is no installed pump.

7.2.1 Sample pumps

Modern groundwater quality monitoring protocols require a large volume of water to be pumped from the borehole *before* the sample is taken to ensure that the sample is representative of the groundwater in the aquifer. Pumped samples are easily obtained from supply wells, or during pumping tests on new wells; otherwise a portable pump is used. This could be either a suction pump (see Section 6.3.1) or a sample pump specially designed for groundwater monitoring. Small-diameter electrical submersible pumps are available from several manufacturers and come with the pump fitted to a length of hose so that it can be easily lowered into the borehole to the required depth (see Figure 7.1). They use a small generator or a mains power supply and are a relatively easy way of purging the borehole before taking a sample. Make sure that you read the manufacturer's instructions before trying it out and follow all the safety rules. It is important to ensure that the pump is sufficiently below the water level to avoid it becoming exposed when the water level draws down. The motor on the end of the pump is cooled by water flowing past it and if it is run without adequate cooling the motor may overheat and burn out. Pumps that run on a 12-V car battery are also available, although the pumping rates and lift are smaller.

In Figure 7.1a a small electric pump has been lowered down the borehole on the left of the picture. The pump is suspended on the yellow flexible-hose rising main which is strong enough to

(a)

(b)

Figure 7.1 *Small electrical submersible pumps can be used for sampling. Photograph (a) shows the control box and delivery hose for a Grundfos MP1 sample pump that is down the borehole. Photograph (b) shows the pump after it has been retrieved from the borehole.*

support its weight, with the power cable strapped to it to reduce the risk of it becoming tangled. Switch gear which is attached to a mains electrical supply stands next to the roll of hose and is used to control the speed of the pump. The pump can also be run using a generator. The pump seen in Figure 7.1b is suitable for use in boreholes with diameters of 50 mm or more and can pump from depths as great as 90 m, delivering up to 2 $m^3 h^{-1}$.

It is possible to take a sample from a specified depth in a borehole while pumping from a shallower depth by inserting a pipe with a diameter big enough to take the pump to the required depth. You must remove a volume equivalent to at least the volume of the pipe before you take the sample.

A simple hand-operated pump that is available in several sizes, one small enough to fit down standard 19-mm-diameter piezometers, is shown in Figure 7.2. The inertial pump simply comprises a simple ball valve on the end of a length of tubing. This foot valve is inserted into the borehole to the pre-selected sample depth and then moved up and down rapidly, with a travel of a few centimetres either by hand or with a powered unit, which is very useful when the groundwater levels are low down. On the down-stroke the ball inside the valve is forced upwards, allowing water to flow into the tube. On the up-stroke the ball is forced downwards, closing the valve. Repetition of this action forces water up the tube until it emerges at the surface. Various sizes are available for borehole

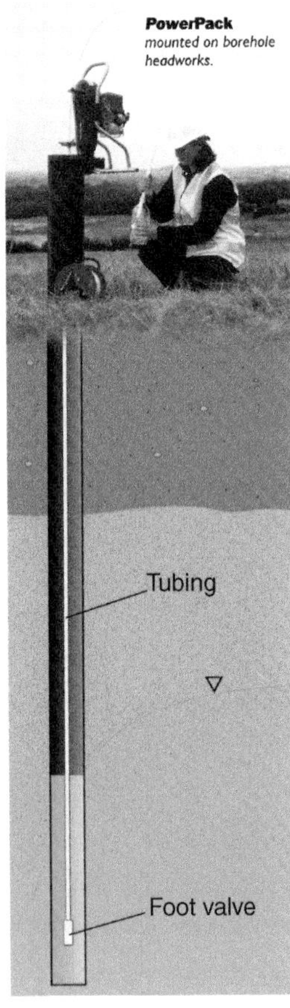

Figure 7.2 *The operator is using an inertial pump driven by a portable generator to operate the pump. (Courtesy of In-Situ Europe Ltd.)*

diameters 15–200 mm. Such pumps are simple to use but harder work than those powered by electricity, unless you use the small portable motor that is strapped to the top of the borehole and moves the pump up and down for you. In the photograph, the operator is using a unit driven by a portable generator to operate the pump.

Sampling groundwater contaminated by volatile organic compounds (VOCs) such as benzene and toluene needs special pumps that prevent the contaminants being lost by evaporation. Bladder pumps are made for this purpose and have a special bag (the bladder) made of inert material such as Teflon that holds and isolates the sample. A valve on the base of the bag allows the sample to enter it and prevents it from flowing out again. Air is used to squeeze the bag from the outside without direct contact with the sample, thereby forcing it up a tube to the surface. When the air pressure is released, more sample flows in through the valve and the process is repeated. The steady flow at small velocities preserves the sample and any VOC it contains.

When sampling from poor-yielding aquifers it may not be possible to purge the borehole before you take the sample. However, it is important to keep a note of the volume of water removed before you took the sample so that it can be taken into account when the results are analysed.

7.2.2 Bailers and depth samplers

The simplest device to obtain a sample from a well or borehole is a bailer consisting of a tube made of stainless steel or plastic, sealed at the bottom and weighted so that it will sink. Alternatively, they may be made of a plastic that is denser than water. You simply drop it down the well on a string and allow it to fill up. There are many companies that manufacture disposable bailers that are designed to be used once or at least a limited number of times. More sophisticated bailers are sealed at the top and have an inlet port on the side just below the top. Depth samplers are more sophisticated still and consist of a stainless-steel or plastic tube, open at each end, which can be closed by spring-loaded bungs, as shown in Figure 7.3. The device is suspended on a steel cable and lowered down a

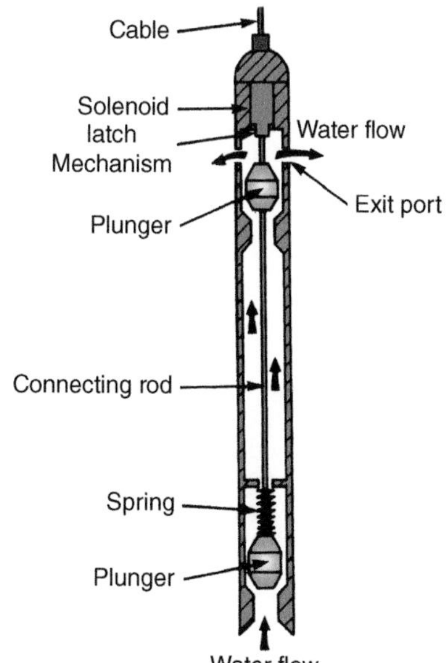

Figure 7.3 *A sample of groundwater can be obtained from a predetermined depth using a depth sampler.*

borehole to the required depth, when the spring mechanism is activated, pulling the two bungs into place. Water continuously flows through the instrument as it is lowered down the borehole, so that when it stops it contains water from that particular depth. The plunger-like valves are closed either electrically (as in Figure 7.3) or by dropping a weight attached to the cable. Electrically operated samplers are usually lowered on the cable that carried the electrical charge. Mechanical samplers are lowered on a small-diameter (2–4 mm) steel-wire cable. Be careful if you are using the latter sort to take a sample from just below the water surface, as it is easy to knock the sampler off the end of the cable. Check that the connections are tight before using it. Bailers and depth samplers come in a range of sizes with capacities from 0.1 to 1 L.

It is sometimes necessary to obtain a sample from a predetermined depth and prevent water from other parts of the borehole entering the sampler. A sampler is available that can be pressurised so that water will not enter it as it is lowered down the borehole (see Figure 7.4). The device consists of a stainless-steel cylinder, some 20 mm diameter, fitted with a pressure-release valve at the base. The top of the cylinder is attached to about 100 m of 5-mm-diameter nylon tube wound on to a drum. The system is pressurised using a tyre pump, with the internal pressure measured by a pressure gauge attached to the drum. In operation, the system is pre-pressurised to a value in excess of the hydrostatic pressure at the depth from which the sample is required (1 bar=10.2 m head of water). Once it has been pressurised, the cylinder is lowered to the selected depth and the air pressure is vented. The relief valve will now open, allowing water to flow into the cylinder. When the sampler is full, the system is pressurised again to close off the release valve and hauled back to the surface.

Before you lower any type of depth sampler into a well, check that it is free from any obstruction and open to its full depth. The easiest way to do this is to lower a total depth probe. If you feel any blockage, it will be necessary to try to free it. A few blows by dropping the depth probe half a metre or so may do the trick. Be careful, however, not to get the depth probe stuck and make the situation worse. You may have to use a CCTV borehole inspection service to check out what is there. Failing that, you may simply resort to taking the sample above the obstruction.

7.2.3 Sampling in awkward locations

It may be necessary to obtain a sample from a seepage or spring flow where it is difficult to insert a bailer or fit a sampling jug beneath the flow. Innovation is important here! You may be able to obtain the sample by using a home-made sample jar that is altered to fit an awkward access or even a polythene

Figure 7.4 *This sampler uses air pressure to control the depth at which water is drawn into the sampler, thereby eliminating possible contamination from water flowing through it as it is lowered down the borehole. (Courtesy of Solinst Canada Ltd.)*

bag. For example, to obtain a sample from a spring flowing along a narrow gully at the bottom of a 2.5 m deep chamber I made an improvised bailer about 15 cm long using the bottom part of a plastic drinks bottle with a diameter of 7.5 cm and small enough to fit into the gully. Stout string was used to make a handle by tying it round grooves in the bottle's sides and made extra secure with electrical insulation tape. Climbing into the chamber was not an option on safety grounds and so the bailer was manipulated into the gully from the surface using three drain rods screwed together. The bailer was secured to one end of the rods using the handle, with the tape used once more for added security. Rods are far more effective than dangling the jar on string in such situations. It is essential, however, to remember to rotate them to the right to tighten the screw joints and avoid them coming apart.

7.2.4 Field measurement of water quality parameters

The usual parameters measured in the field include pH, Eh, dissolved oxygen, temperature and conductivity. (Eh is the redox potential or oxidation/reduction potential.) Some sampling protocols stipulate that a number of parameters such as pH and Eh should be continuously monitored during purging and the pumping extended until they are stable. Wellhead testing also helps you assess how much change has occurred between the sample being taken and the laboratory analysis being carried out. Such changes are usually caused by the large reduction in the pressure of the groundwater sample as it was removed from the aquifer and contact with atmospheric oxygen.

A range of battery-operated instruments is available that will enable you to take these measurements and those of many other parameters very easily. It is usual to set up several in-line test cells, as shown in Figure 7.5, each of which contains an electronic probe. On-line sampling is used to take water samples without the sample coming into contact with the atmosphere. Water from the pumping main is directed through a hose into a sealed jar that has ion-specific probes fitted into it so that the values can be read from the meters. Water is allowed to flow through the system for several minutes, with care taken to ensure that the jars are completely full of water. The instruments are monitored until constant values for the parameters are obtained. Water from the sealed jar is passed into the bucket that contains sample bottles for tritium and CFCs.

Monitoring devices comprising ion-specific probes or sensors for pH, Eh, or temperature attached to a data logger can be installed in boreholes to record groundwater quality, often monitoring several parameters at the same time. Such systems are used in pollution studies and to provide an early warning of groundwater quality changes that may affect water supply sources.

On-site measurements may be your only option in obtaining chemistry data in areas where laboratory facilities are not available or where results are required quickly. Water quality parameters measured in this way can provide sufficient information to assess the value of the local groundwater for drinking water supplies or determine the movement of specific contaminants. Remember, however, that field measurements are not a substitute for laboratory tests.

Portable instruments for pH and conductivity have sensors that are inserted into a water sample and the value read off the digital display. Indicator paper can also be used to measure pH, although this does not give such accurate results. Electronic thermometers have a probe, sometimes at the end of a cable, that is inserted into the water and the temperature read off the display. All these instruments need calibration before they are taken into the field. Follow the manufacturer's instructions and keep a diary to demonstrate when the instrument was last recalibrated.

The electrical conductivity of water changes with its temperature and consequently measurements are usually recorded for a standard temperature of either 20°C (in general use) or 25°C. Most instruments correct for temperature automatically, but it is essential to note which reference temperature has been used. If a measurement has been made at the ambient temperature it needs to be corrected to the appropriate temperature. The electrical conductivity of a typical groundwater sample varies by about 2% per degree Centigrade change in temperature. The relationship is not linear and also varies with the concentration of minerals in the solution. Table 7.2 gives the relative

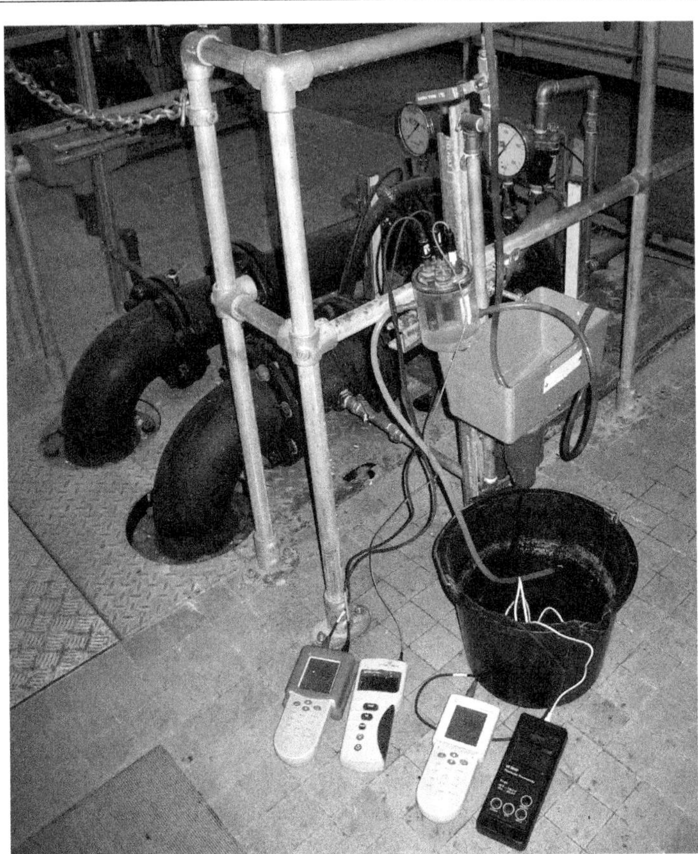

Figure 7.5 *On-site measurements of temperature, specific electrical conductance (SEC), alkalinity, pH, dissolved oxygen (DO) and redox potential (Eh) were taken at a water utility sampling point in the Chalk in Hampshire. Alkalinity was measured by titration against* H_2SO_4 *and was reported as* HCO_3. *The pH, Eh, and DO were measured in a flow cell to restrict aeration and were monitored until stable readings were obtained. (Courtesy of the Environment Agency and the British Geological Survey. © NERC. All rights reserved.)*

Table 7.2 *Relationship between conductivity of solutions at different temperatures to a standard of 20°C.*

Solution	5°C	10°C	15°C	20°C	25°C
NaCl	0.689	0.788	0.891	1.000	1.112
KCl	0.693	0.815	0.907	1.000	1.095

See text for use of this table.

value of the conductivity of sodium and potassium chloride solutions with respect to a standard temperature of 20°C. You can use these values to calculate an approximate conductivity value for solutions at different temperatures. The relationship for each of the two salts is slightly different, although, when taken to the first decimal place, the values are the same. Even if the solution is not dominated by either of these salts, these conversion factors will give you an approximate value for the conductivity at the temperature you need.

7.2.5 Sampling for dissolved gases

In sampling for dissolved gases such as oxygen, carbon dioxide, nitrogen, and methane, the water sample can be taken using a low-pressure system that can also be used for taking groundwater samples to avoid contact with the atmosphere, as well as sampling for a wide range of dissolved gases. The example shown in Figure 7.6 is one of the specially designed systems for sample collection for dissolved gas analysis and comprises a bag (or flask) that is supplied flat and devoid of air. It does contain a bactericide capsule to prevent bacterial degradation of the sample. It is easy to fill and each bag comes with a sample tube for a one-off use. Such systems are provided with instructions that are simple to follow.

In sampling for noble gases that have a low solubility concentration, it is important to maintain the absolute pressure of the groundwater sample to prevent de-gassing, and a special rig will be required, such as the one shown in Figure 7.7. Gas solubility is dependent on the water pressure, which is determined by the depth from which the sample is obtained. It is important therefore to preserve the sample at a high pressure.

In Figure 7.7 the general arrangement for taking a sample is shown to ensure that the sample pressure is maintained at 3 bar. The pipe from the sample pump is fitted with a T-piece, with flow controlled by two valves, with the sample line having a pressure gauge and a valve to ensure that adequate pressure is maintained. The sample is captured in a length of copper pipe (approximately 6mm internal diameter and some 30–50 cm in length) to provide an adequate sample volume for analysis. The copper pipe is fitted into a special assembly (Figure 7.7a) comprising two clamps on

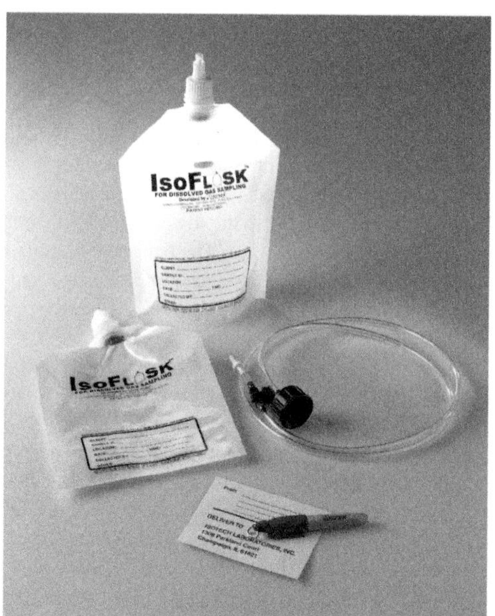

Figure 7.6 *The IsoFlask sample bag comes empty except for a biocide. It is filled using the one-off-use sample tube. (Courtesy of Isotech Laboratories, Inc. (A Weatherford Company), Champaign, Illinois, USA.)*

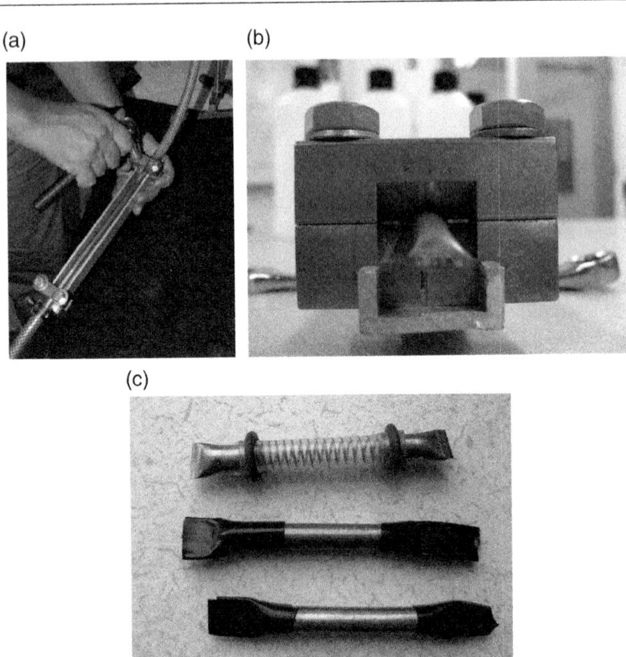

Figure 7.7 *This series of photographs illustrates the way that noble gases are sampled. Water is passed through the small-diameter copper tube in (a) and then clamps (b) at each end are fastened, with the one at the discharge end being first. The sample is preserved in the tube at a high pressure (>3 bar) as in (c). (Courtesy of the Department of Geology and Geophysics, University of Utah, Salt Lake City, Utah, USA.)*

a base-plate, and connected to the feed and discharge pipes using screw-operated pipe clamps (jubilee clips). Each clamp is made of two short, square-section metal bars that are held together by a pair of bolts (Figure 7.7b). The bolts are fixed to the lower bar so that the upper bar can be firmly clamped against the lower one using hexagonal nuts. The bars have a v-shaped middle section with a rounded edge so that it will not cut into the copper pipe, with the gap between the rounded v-shaped edges being slightly less than the wall thickness of the copper pipe so that they seal the pipe by crushing it (or swaging it) and cutting is avoided (Figure 7.7c).

The sample is taken by setting up the equipment and adjusting the valves until the flow through the sample line has a pressure gauge reading of at least 3 bars. It is vital to avoid bubble formation in the sample tube as this will affect the analytical result. Bubbles can form where the sample is taken from depth and the pressure difference at the surface is significant. Tap the pipe as the water is flowing through it to dislodge any bubbles that may have formed, and check the transparent pipe for signs of bubbles. If necessary, partly close the valve at the end of the system gently, which will usually eliminate bubble formation. The nuts on the clamp at the end of the system are tightened until the flow is cut off, quickly followed by tightening the other clamp, thereby sealing the copper pipe and capturing the sample with all the dissolved gases in solution. The sample tubes should be preserved with the ends wrapped in electrical tape and transported to the laboratory for GCMS analysis. The best pumps for dissolved gas sampling are electrical submersibles that can be regulated to produce a low flow, such as the Grundfos MP1 or similar. The use of bladder pumps is often a problem, especially if they employ a Teflon bladder as this material is permeable to gases, especially helium.

The atmospheric trace gases CFC-11, CFC-12, CFC-113 and SF6 (sulphur hexafluoride) are increasingly being used as tracers of groundwater residence time. Large-scale production of CFC-12 began in the early 1940s, followed by CFC-11 in the 1950s and CFC-113 in the 1960s. CFC-11 and CFC-12 were used mainly for refrigeration and air-conditioning, while CFC-113 was used as a solvent. Inevitably they leaked into the environment, with atmospheric concentrations rising until the 1990s, when production was cut back to protect the ozone layer. SF6, another industry-derived gas, has been detectable in the atmosphere since the early 1960s and is still rising strongly in concentration.

To sample for CFCs you will need special 1-L bottles with screw tops that are supplied by the laboratory, a 10-L bucket or a 4-L beaker (plastic is fine), and refrigeration-grade copper tubing, as plastic hoses may leach into the sample and can affect the analysis. Have enough bottles to take four samples per site. The bottles and caps should be thoroughly flushed with the sample water, with the bottles being both filled and capped under water.

First ensure that the borehole is purged – the connecting copper tubing should be well flushed through (see Table 7.8) – and that the pumping rate is ideally between 30 and 240 L h^{-1}. Place the bottle inside the empty bucket and insert the copper tubing from the wellhead into the bottom of the bottle. Allow the bottle to fill and overflow into the bucket, which should be allowed to fill and eventually overflow; a minimum of 1 L of water should have passed through the bottle, but remember that flushing the bottle with more water is far better than with not enough water. Place the bottle cap under water with the foil side facing upwards and tap it to dislodge air bubbles, or use the flow from the copper tubing inlet line to remove any bubbles by pointing it into the cap while under water. As filling takes place, release any air bubbles trapped in the bottle and remove the tubing when the bucket overflows. The flow through the tube should be slow enough to be laminar to reduce the chance of gases being dissolved from the atmosphere. Figure 7.8 shows the procedure in pictures.

Keep the water level at the top of the bucket to provide a seal against air contamination and screw the cap onto the bottle, making sure that it remains below the water level in the bucket so that the water in it does not come into contact with the atmosphere. The cold groundwater may make your hands and forearms uncomfortable, so be warned!

Remove the bottle from the bucket, make sure that the cap is fastened tightly and dry it off. Then invert it and look for bubbles; if there are any, the process needs to be repeated from the beginning, but use a new cap for the bottle. Assuming there are no bubbles, the sampling is complete. Tape the cap securely to the bottle with electrical tape; wrap the tape in a clockwise direction, looking down from the bottle top. Three rounds of electrical tape are needed to keep the cap sealed. Label each bottle with the name of the well, date, time of sampling and the sequence number of each bottle as it was collected. It is recommended to store and ship the bottles upside down at room temperature. It is ideal to use the original cardboard box they were delivered in. Make sure that they do not warm up above room temperature (about 23°C) and if a bubble forms in the sample while in storage it is normal. When transporting the samples to the laboratory for CFC analysis, use shredded paper as packing; avoid packing materials such as polystyrene chips and bubble-wrap which can be sources of CFC contamination.

The filling and capping procedure for SF6 is done in the open air; do not submerge the bottle and cap in a bucket for filling and capping. It is important that water does not enter the area behind the cone seal in the SF6 bottle cap because this area behind the cone seal allows the water to expand as it warms up to room temperature without breaking the bottle.

The well should be purged as before. Place the tubing into the bottom of the 1-L bottle and allow it to fill and overflow from the neck. At least 3 L of the sample water should discharge from the bottle with the tubing at the bottom of the bottle; then slowly remove the tubing while the water still flows. Cap the bottle without leaving any headspace and tape the cap with electrical tape as before. Collect two bottles per site, labelling and storing as with the CFC samples. The samples can be kept at room temperature but should not be allowed to get any warmer.

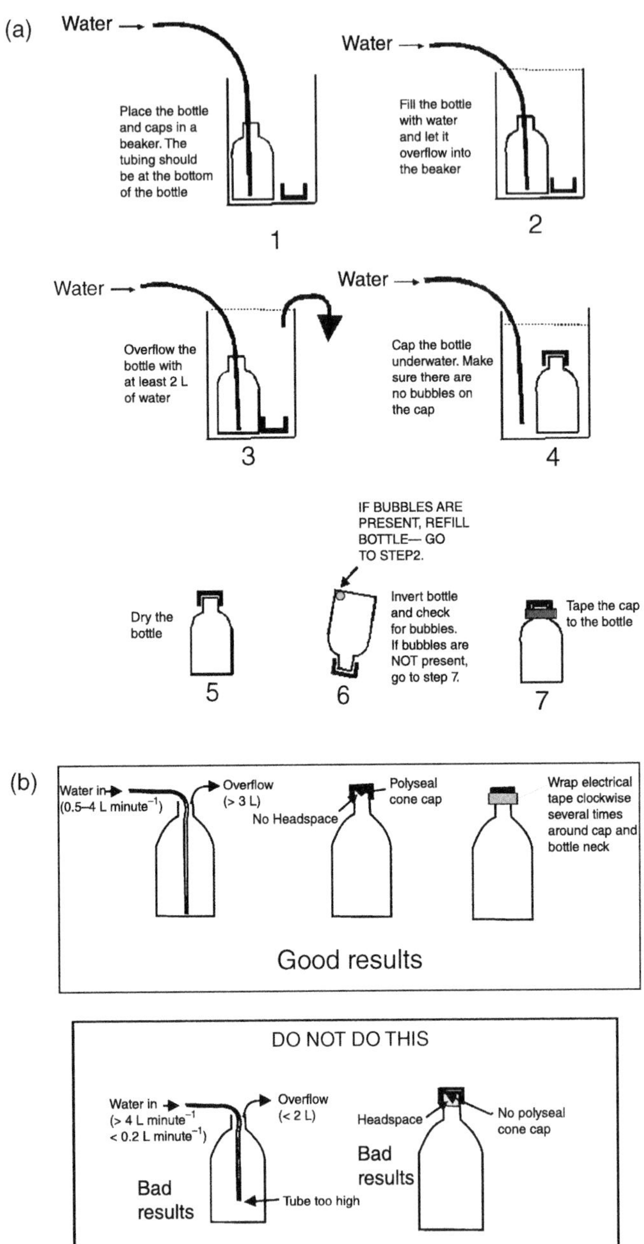

Figure 7.8 *The method of sampling for CFCs is shown in (a) and for SF6 in(b). Note the same equipment can be used for both samples, although only CFCs need to be sampled underwater. (Figures from the USGS website with permission.)*

7.3 Sampling Protocols

The way in which you take groundwater samples will have a bearing on the validity of any conclusions you reach based on their chemical analysis. It is important to follow standard procedures (or *protocols*) that are designed to ensure that the water chemistry is not changed by the sampling process. This means that you can compare the results from one site with another with a high level of confidence. The protocol will cover all aspects of taking the sample, including sample storage and transport to the laboratory, as well as the use of the equipment and on-site measurements.

Most environmental regulators are likely to specify their own sampling protocols, so make sure that you obtain a copy and follow them. It is equally important to check with the laboratory staff for details of any procedures they require to ensure that the sample is in the best condition possible when it arrives for analysis. The protocols given here conform to those used by the Environment Agency in England and are likely to be acceptable to most regulators anywhere.

7.3.1 Pre-planning

Plan your fieldwork in detail before setting out from the office. List the sampling sites with details of the sources and decide on the best order to take the samples based on travelling considerations between sites. Take account of the amount of time needed to purge the well before each sample is taken to work out how many sites you can visit in one day. Decide on the details of the chemical and bacteriological analyses you need. Use the checklist provided in Table 7.3 for your planning activity, and the equipment checklist in Table 7.4 to make sure that you have everything with you when you set out. Examine the equipment in the comfort of your office and read the manufacturer's instructions so that you are able to operate any new equipment *before* you go into the field. Make sure that it is all working before you set off, and that bottles, pipes, and sampling equipment are all clean.

Despite taking great care, it is still possible for samples to be polluted during the sampling process from contaminants circulating in the air. Examples include engine exhaust fumes (PAHs), crop sprays and even microbes coughed out by the sampler! If you have unexpected contaminants in the analytical results, especially with low concentrations, do not panic! Think about how the water could have become contaminated during the sampling process and then repeat it with care, taking appropriate precautions. When a sample is polluted you usually find that the contamination is limited to one or two PAHs or pesticide species or low values of total viable counts (TVCs), whereas you normally get a much wider range when the groundwater itself is contaminated. Other clues include the sampling location, such as near an open door with vehicles moving outside or from a borehole next to a diesel-powered generator or just by a busy main road.

Table 7.3 *Checklist to prepare for groundwater sampling.*

1. Prepare list of wells and springs you intend to sample, including addresses, map references and maps of the area. Plan route and work out approximate timings for arriving at each site.
2. Check with owners that you can have access at the planned time.
3. Prepare a list of determinands, using Table 7.1 as a guide.
4. Arrange for the laboratory to provide appropriate sample bottles, with coded labels where appropriate. Make sure laboratory codes match the type of analyses you require. Take spare bottles in case of damage or need for extra samples.
5. Arrange for temporary sample storage and courier services if needed.
6. Place empty bottles in cold boxes for transport, with ice packs when needed, and some form of packing such as bubble-wrap.
7. Prepare field sheets with details of each sample site.
8. Assemble equipment you will need (see Table 7.4).
9. Check your safety equipment and ensure you follow appropriate safety procedures (see Appendix I).

Table 7.4 Checklist for sampling equipment.

1. List of sample sites and type of sample to be taken at each location.
2. Route map and site information such as address, name of owner, telephone number, and so on.
3. Safety equipment such as first-aid box, weatherproof clothing, safety hat, gloves, spare clothes and so on.
4. Appropriate bottles for all samples, with several spares, packed in cold boxes with ice packs and suitable packing material.
5. Flow-through cells, including all necessary pipes and clips with spares.
   Glass beakers for temporary sample storage before on-site measurement. Meters with probes to measure temperature, pH, Eh, electrical conductivity, and dissolved oxygen. Ensure that probes fit flow-through cells and you have calibration solutions available on site.
6. Alkalinity titration equipment, with adequate supply of cartridges and spares.
7. Cleaning solutions for instruments (where needed) and adequate supply of de-ionized water in a dedicated jerry can. Supply of cleaning tissues.
8. 'Blank' samples of water of known quality to test accuracy of laboratory or field procedures.
9. Bailer, bucket, jug, or glass beaker for spring sampling, as appropriate.
   May require plastic bag for small samples. Wash in cleaning solution and rinse well in de-ionized water. Spring sampling may also require use of a small 12-volt suction pump. Ensure that both pump and ancillary pipes are clean.
10. Depth-samplers or bailers, if appropriate. Check they are clean and that depth-sampler valves are working. Check cable to ensure it is long enough to reach deepest sample point; ensure batteries are charged; and that firing mechanism works.
11. Inertial pump set, if appropriate. Check foot valve(s) and that there is adequate tubing. Make sure that pump and tubing are clean, using cleaning solution and rinsing as required.
12. Electrical submersible sample pump, if appropriate. Check that generator is working and cable connections are in good order. Make sure that pump and hose are clean. Check you have an adequate supply of fuel and the appropriate cans and funnels.
13. In-line filters and/or filter paper, with adequate spares.
14. Plastic sheeting, enough rubber gloves, torch, dipper, total depth probe, tool-box, bucket, electrician's tape, penknife, hose, spare pencils, pens, and notebook. Do not forget keys to locks on borehole headworks.

7.3.2 Field procedures

Tables 7.5–7.9 provide a step-by-step guide to each stage of taking samples, from the range of different situations you will face. There are a number of common elements in all groundwater sampling, such as avoiding contamination of the sample and correct sample storage. Do not forget to keep the equipment clean by laying it on plastic sheeting or in a specialist container rather than on the ground. You should always take wellhead measurements using flow-through cells if possible. Samples taken from a spring in a jug or bucket, and depth samples from a well, cannot be used to reliably measure Eh or dissolved oxygen.

Groundwater samples should be filtered unless alternative special procedures are required because of the parameters you are examining. Filtering does matter: I was once involved in a job where some samples had been taken from boreholes drilled in a limestone aquifer. The results came back with calcium in very high concentrations and it was only when I checked with the laboratory that I discovered that the analytical method used took account of all the suspended matter in the samples which were not filtered. When I checked back with the samplers they confirmed that they had not filtered the samples and that they looked cloudy!

Use in-line filters so the sample pump forces the water through the filter. Where a sample is obtained from a spring or from a borehole with a bailer it is necessary to pump the water through the filter using either a hand-operated rig or a small 12-volt centrifugal pump. If necessary, a funnel

Table 7.5 Protocol for sampling from springs.

1. Put on appropriate safety equipment: for example, rubber gloves to avoid contamination of the source; safety harness and line or life jacket if size of catch pit warrants it.
2. Remove cover from catch pit and check for debris, making sure that none falls in. Clean out catch pit and allow it to clear. Identify in-flow point(s) into catch pit and try to take the sample as close to this point as possible. Lay out sampling equipment on plastic sheet next to the spring.
3. If using a jug, bucket, or bailer to take the sample, first install electronic probes to measure conductivity and temperature at in-flow point. Monitor readings and wait until they have stabilised. Record meter readings and then take the sample using the jug, bucket, or bailer. Pour sample into filter equipment and fill sample bottles as required.
4. If the 12-volt sample pump is to be used, position suction pipe at an appropriate place in catch pit near in-flow. Make sure suction does not disturb any particles on catch pit floor. Connect pump discharge pipe to flow-through cells and run pump for a few minutes to flush cells. Switch off and insert electronic probes. Run pump until instruments equilibrate, then record readings for temperature, pH, electrical conductivity, Eh, and dissolved oxygen. Take a sample in a clean glass beaker for alkalinity determination.
5. Stop pump and connect discharge pipe to filter equipment. Fill sample bottles. Label bottles and place them in cold boxes, ensuring they do not lie in direct contact with ice packs by using packaging material to separate them.
6. Clean equipment and rinse with de-ionized water ready for sampling at next site.

Table 7.6 Protocol for sampling from a well using depth samplers.

1. Check borehole records in the office to determine casing details and information from geophysical logging to determine in-flow zones to decide on number and depths of samples.
2. Open borehole ready for sampling and lay out equipment on plastic sheet. Measure depth to water and check borehole is unobstructed, using total depth probe.
3. Connect the various parts of the depth sampler and make sure you can read depths on the cable for each sample point. Rinse sampler three times in de-ionized water.
4. Having made sure that ports are open, steadily lower sampler into borehole until requisite depth has been reached. Activate closing mechanism and lift sampler back to the surface.
5. Empty sampler slowly into a clean glass beaker for on-site measurements with probes. Pour remainder into filtration equipment. Repeat steps 4 and 5 until an adequate volume of water has been obtained.
6. Take a sub-sample from water in glass beaker and make an alkalinity measurement. Measure temperature, pH and conductivity of remaining sample. (Note: Eh and dissolved oxygen cannot be measured reliably.)
7. Filter water in filter equipment and fill appropriate bottles. Label bottles and place them in cold boxes, ensuring they do not lie in direct contact with ice packs.
8. Rinse equipment with de-ionized water and repeat steps 4–7 for each depth where a sample is required.
9. Clean equipment and rinse in de-ionized water ready for sampling at next site.

lined with filter paper (or even coffee filters at a push) may be used, although this method has the disadvantage that the oxygen content of the water sample is increased as the water is poured through the filter and that could result in reactions causing precipitation of some metals.

Samples for bacterial analyses require special care. Disinfect the sample tap with a chlorine-based solution or peracetic acid, which is used as a disinfectant in the food industry, or use a blow torch and flame the tap for several minutes to destroy any microbes on it. Do not flame the tap if you are taking samples for PAH determination, take the PAH sample first. Take care not to burn yourself with either the strong disinfectants or the blow torch. When you take the sample, hold your breath to make sure that you do not breathe on it.

Table 7.7 *Protocol for sampling from a well using an installed pump.*

1. If not already running, start pump and allow it to run for at least 15 minutes before taking a sample. If pump is not running frequently you will need to purge-pump to remove three-times the water standing in the borehole before a sample is taken (see text and Table 7.8).
2. Identify sample tap you will use and set out your equipment nearby, laying it on plastic sheet to keep clean.
3. Fit sampling hose to tap (if needed), place bucket underneath and run tap. Use water to rinse tap and surrounding area. Run tap to waste for at least 10 minutes. Use a longer length of hose if necessary to avoid flooding round sample tap.
4. Switch off tap and connect to flow-through cells. Fit electronic probes and open tap. Wait until instruments have settled down before recording readings on your field sheet. Leave tap running to take a small sample to titrate for alkalinity.
5. Switch off tap and disconnect flow-through cell. Fit filter equipment to hose and open tap once more. Fill appropriate sample bottles, label and place in cold boxes, ensuring they do not lie in direct contact with ice packs.
6. Switch off tap and disconnect filter unit.
7. Clean equipment and rinse in de-ionized water ready for sampling at next site.

Table 7.8 *Protocol for sampling from a well using an electrical submersible sample pump.*

1. Before going to site, use borehole records to calculate volume of water contained in unpumped borehole. Then from pump performance characteristics calculate how long it will take to purge-pump at least three times this volume.
2. Remove borehole cover to allow access for pump and set out equipment nearby, laying it on plastic sheet to keep clean. It is essential to avoid the pump, rising main, hoses, and electrical cable coming into contact with the ground and being contaminated with dirt, grass, and so on. Where appropriate, set up tripod above borehole.
3. Measure standing water level in well and use this value to calculate volume of water contained in unpumped borehole. From pump performance characteristics calculate how long it will take to purge-pump at least three times this volume. Use total depth probe to check depth and for obstructions.
4. Assemble pumping equipment including generator. Where rising main and electric cable are separate, tape or tie them together to prevent them from tangling in borehole. Use electrician's tape as it is easily broken once pump has been removed. If using a separate purge pump, tape or tie the two pumps and rising main assemblies together so that sample pump is about 1 m below purge pump.
5. Calculate length of rising main that will have to be lowered into borehole to place bottom of sample pump about 2 m above bottom of borehole. Mark rising main at position where it will lie at level of borehole top when pump is in place.
6. Lower pumping assembly into the borehole using the winch to control weight until mark on rising main is level with well top. Connect discharge hose to purge pump (or sample pump if this is being used for purging) and lead it to a convenient place to discharge purge water. Ensure well head will not become wet.
7. Start generator and commence purge pumping. Note time you started pumping and check discharge rate using stopwatch and bucket to verify calculated time to purge well. Once this time has elapsed, shut off purge pump and start sample pump (or commence sampling procedure if using the same pump).
8. Fit hose to sample pump and connect to flow-through cells using a valve to control flow of water. Fit electronic probes and start pump. Once there is a flow of water, open valve and divert flow through cells. Divert over-flow away from wellhead.
9. Follow measurement and sampling procedure described in steps 4 and 5 in Table 7.7.
10. Switch off tap and disconnect filter unit. Rinse equipment in de-ionized water three times in preparation for next site.
11. Withdraw pump assembly, avoiding tangling of pipes and cables. Disconnect various parts, taking care to keep them clean. Rinse sample pump with de-ionized water.

Table 7.9 Protocol for sampling from a well using an inertial pump.

1. Follow procedures described in steps 1–3 in Table 7.8.
2. Assemble inertial pump using sufficient tubing for borehole depth. Using total depth measurement, mark tube with tape at a point where pump suction will be about 2 m above bottom of borehole in operation.
3. Insert pump into borehole, holding it securely. Operate pump until flow of water starts. Discharge water into bucket to measure volume removed. Continue until three times the volume of water in well has been purged. If necessary empty bucket away from well.
4. Follow procedures described in steps 8–11 in Table 7.8.

7.3.3 Sample bottles

It is essential to have a supply of clean bottles for your samples for transport back to the laboratory. Most laboratories will provide bottles of an appropriate size, cleaned ready for use, and sometimes with the labels already printed! Otherwise, check with the laboratory how much water they require before setting out, and ensure that you have an adequate supply of the right size bottles with you. If in doubt, use glass bottles as plastic ones affect some types of analysis. Laboratories usually need the sample divided into half-a-dozen bottles or more because different parts of the analysis will be done in different locations.

If you need to take some samples but do not have any sample bottles with you and have not got time to go to the laboratory to pick some up, you can improvise by buying sealed glass bottles of mineral or spring water and then emptying them out. Give them a shake to make sure that they are emptied completely and rinse the bottle out with the water you are sampling before filling it with the actual sample. Provided the bottles are sealed they should be sterile, although if you chose sparkling (i.e. fizzy) water, the dissolved carbon dioxide will act as a mild biocide and provides extra reassurances that the bottles are sterile.

Use a hose or funnel to avoid spillage, and rinse the hose, funnel, bottle, and bottle cap with the water you are sampling. Fill the bottle right up so that the water meniscus is above the top, fill the cap with water, then turn it over quickly and screw it on. With practice this method ensures that the bottle is completely full. Groundwater samples often have a small 'air' bubble in them. This is gas (usually CO_2 or N) that has come out of solution as its saturation concentration adjusts to the reduced pressure conditions compared with those in the aquifer. The pressure is reduced by one atmosphere for just over every 10 m depth below the well water level to the sample point. Such differences can be significant, so the act of taking the sample starts a process of changing the chemistry of the sample! This is why it is very important both to take on-site measurements and to store the sample correctly. Oxygen in the atmosphere will react with most metals in solution, causing them to precipitate as insoluble oxides and hydroxides. To overcome this problem, about 50 mL of water should be placed in a separate bottle that contains 1 mL of concentrated nitric acid (65% HNO_3), to keep the metals in solution. Specially prepared bottles will be provided by the laboratory, so do not rinse them and remove the acid! It is advisable to specify the analysis for both iron and manganese to be measured as both the total and dissolved concentrations. Once these dissolved metals come into contact with oxygen in the atmosphere they start to form oxides, which are insoluble and will precipitate in the sample bottle, making the dissolved value too low. On the other hand, any rust particles from the casing or pipework will be dissolved in the acid and give artificially high results for the iron. If you are sampling from a pipe that has been cut to install a sample tap, be aware that a minute particle of swarf may get into the sample and significantly increase the apparent copper or lead content of the water. The best way to avoid this problem is to filter the water before it is put into the bottle.

Label all the bottles very clearly, identifying the sample location, the depth from which the sample was taken and to indicate to the laboratory the analysis you require on that sample. Write the label *before* you fill the bottle, as it is very difficult to write on wet paper! Make a record of each sample in your field notebook and include the results of any field readings. A checklist of the information that should be recorded at each site is provided in Table 7.10.

Take great care not to drop the full bottles. Wet glass and plastic are slippery and you do not want to have to repeat depth sampling if it can be avoided – it's hard work! Transport the sample to the laboratory without delay so that significant changes do not occur before analysis. Use an insulated cool box to transport your samples rather than a bottle crate or holder. Take several frozen ice packs to make sure that the temperature in the cool box stays under 4°C. Use bubble-wrap or even screwed-up paper to keep the bottles apart and prevent the ice packs from touching any of the bottles. If you need to keep the samples overnight, store them in a refrigerator, but make sure that they do not freeze. Many laboratory companies have drop-off points where you can leave your samples often in a refrigerator. The samples are taken to the laboratory at regular times, making your life much easier. Contact the company to see where the nearest drop-off point is to your sampling location and the time when the samples are collected. Where these services do not exist, consider using a courier firm to send the samples to the laboratory in a refrigerated van if the distances are too large for you to drop them off. Most laboratories offer this service to their customers.

You can expect the laboratory company to have its own forms that also need to be completed. Either get hold of copies of blank forms that you can fill out with both details of the samples and the analyses that are required or make sure that you fill out the forms when you drop the samples off at the laboratory or the sample point. Most laboratories handle very large numbers of samples every day and accurate labelling and recording is essential to avoid someone else's samples being mixed up with yours.

Table 7.10 *Information to be recorded in the field for groundwater samples.*

General information	
Date:	day/month/year.
Time:	hours/minutes (indicate time zone, daylight saving, etc.).
Sampler:	name and details.
Sample point:	location of where sample was taken (e.g. Brown's No. 2 borehole or Clegg's spring).
Description:	comment on any relevant aspect of the sample point condition.
Sample point number:	reference number in well catalogue, and so on.
Total depth of borehole:	value in metres.
Rest/pumping water level:	note whether pumping or not. Give details of the datum (e.g. top of casing or ground level) and give depth in metres below this datum.
Sampler's comments:	note any relevant information to help interpret the data. Describe the general appearance of the sample (e.g. clear, cloudy, coloured (state colour)).
Field measurements	
Depth sample:	give depth below datum and specify datum used.
Pumped sample:	state how long the pump was running before the sample was taken.
Temperature:	value in degrees Centigrade.
Dissolved oxygen (DO):	value either as DO percentage saturation or in milligrams per litre.
Alkalinity:	test kit adjusts to pH 4.5 and gives values in milligrams per litre as $CaCO_3$. See Table 7.13 to convert units.
Eh:	value in millivolts.
pH:	value in pH units.
Electrical conductivity (EC):	value in micro-siemens per centimetre measured at 25°C.

If you are taking samples from several sources for bacterial analysis isolate the sample bottles from each source by placing them in a plastic bag and sealing it with a tie. This will avoid water from one source coming into contact with the bottles from another and reduces the chance of cross-contamination.

It is very important to observe the safety rules when you are taking and transporting samples from a waste disposal site, or from where the groundwater could be contaminated. Wear rubber gloves and goggles to prevent getting any sample on your skin or in your eyes. Ensure that the sample cannot spill in your car, and avoid having samples that may contain volatile substances in an enclosed space such as a vehicle. These fumes could poison you! Clean all the equipment with special cleaner sold for the purpose and rinse well with clean water. *Do not use this equipment in boreholes used for a water supply.*

7.3.4 Using BART test kits

A number of bacterial species can be found in groundwater and can cause problems in both boreholes and the water quality. A number can be detected using the biological activity reaction test (BART) that consists of a special test tube that has been pre-seeded with a dry culture medium designed to encourage a particular group of bacteria to develop. The concentration of these bacteria is determined by the time lag in days before the reaction is observed.

There are nine different tests identified by the coloured cap coding and the initial letters preceding the word BART and are made by Droycon Bioconcepts Inc. of Regina, Saskatchewan in Canada (http://www.dbi.ca/BARTs). The most commonly used ones are Iron Related Bacteria (IRB-BART – red cap), Sulphate Reducing Bacteria (SRB-BART – black cap), and Slime Forming Bacteria (SLYM-BART – green cap), as shown in Figure 7.9. You can download a user manual from their website https://www.dbi.ca/BARTs/FAQ.html) as well as quick break training guides that have useful tips.

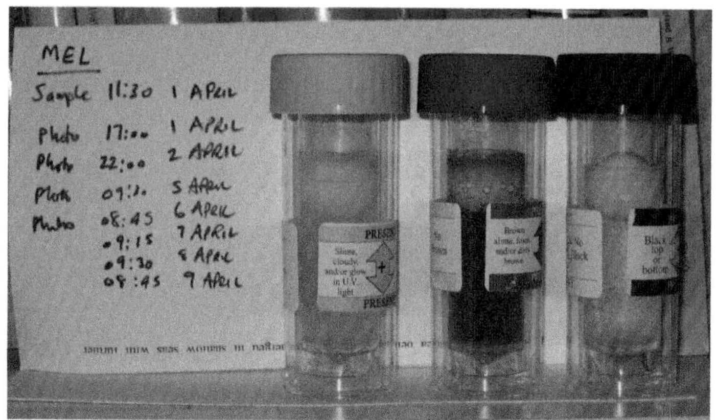

Figure 7.9 *The most commonly used BART test kits are Iron Related Bacteria (IRB-BART – red top), Sulphate Reducing Bacteria (SRB-BART – black top) and Slime Forming Bacteria (SLYM-BART – green top). The samples in the picture were observed for 8 days. The IRB-BART shows aggressive iron-related bacteria to be present; SLYM-BART shows the presence of slime-forming bacteria; and SRB-BART shows little sulphate-reducing bacteria, although the cloudy solution indicates the presence of anaerobic bacteria.*

The BART test kit comes as a double test tube. Simply remove the top off the outer tube and take the inner tube out. Fill the outer tube with a sample and then gently pour this into the inner tube, taking care to fill it to the pre-marked line. Then put the top back on the inner tube, taking care not to touch the inside of the cap or the tube, and then put the inner tube back into the outer tube when you have emptied any remaining sample, and put the cap back on. Label the top of the cap with the sample date and origin. Place the BART away from direct sunlight and allow it to incubate at room temperature.

To check how the test is progressing, each day lift the inner test vial carefully out of the outer BART test tube and view through the inner tube against an indirect light. A common list of the methodologies and applications would be as follows:

- IRB-BART: test is positive when foam is produced and/or a brown colour develops as a ring or dirty solution. The time lag to that event is the delay. A negative has no brown colour developing, no foaming or clouding. This test is commonly used to detect plugging, corrosion, cloudiness and colour. The bacteria that may be detected by this test include iron-oxidising and iron-reducing bacteria, the sheathed iron bacteria *Gallionella*, and pseudomonad and enteric bacteria.
- SRB-BART: a positive test occurs when there is a blackening either in the base cone of the inner test vial (80% of the time) or around the ball (20% of the time). The culture medium is specific for the SRB, such as *Desulfovibrio* or *Desulfotomaculum*. This is a more specific test and specifically relates to corrosion problems, taste and odour problems ('rotten egg' odours), and blackened waters. Slimes rich in SRB tend to also be black in colour. A negative indication occurs when there is an absence of blackening in the base cone of the inner test vial or around the ball.
- SLYM-BART: some bacteria can produce copious amounts of slime that can contribute to plugging, loss in efficiency of heat exchangers, clouding, and taste and odour problems. This is one of the most sensitive BART tests. A positive test involves a cloudy reaction in the inner test vial, often with thick gel-like rings around the ball. A negative test remains clear.

7.3.5 Using borehole packers

Packers can be used to take groundwater samples from selected depths in the aquifer. Set up your packer equipment in the same way as for a pumping test (as an example see Figure 6.16). You should then follow the protocol outlined in Table 7.8 to take the sample, making sure that you adequately purge the system first.

7.4 Monitoring Networks

Groundwater quality monitoring networks are needed to identify trends caused by seasonal variations, the impact of abstraction on seawater intrusion in coastal areas, and pollution problems. As flow occurs in open boreholes, mixing water from different depths, monitoring networks are usually based on pumped boreholes that provide a mixed (or 'average') sample of the groundwater.

From a practical point of view, any monitoring network you establish in your study area is likely to be limited to the availability of sample points. Your budget may well limit the number of analyses, so choose sample points to give the best cover. Try to select wells that provide an even geographical spread and also take account of the different aquifers. Sampling frequency can be variable, ranging from once a month to once a year for general monitoring. For special studies, however, more frequent sampling is often carried out.

If you are studying a landfill or contaminated land, you will need a network round the site and probably across it. These are usually shallow boreholes with a narrow diameter and may be equipped with passive samplers or, in the case of landfill sites, may also be used for gas monitoring.

7.5 Using Chemical Data

Groundwater chemistry defines its suitability for water supply use or to answer straightforward questions on contamination. One of the most common questions will be 'Is the water from the new source potable?' The information in Table 7.11 will help you find the answer and is based on the European drinking water standards, which are broadly the same as those applied in the rest of the world. In hydrogeological investigations, however, groundwater chemistry can be used to extend your understanding of the groundwater system as a whole. For example, water chemistry can provide a valuable clue as to the nature of springs. By comparing the chemistry of different spring waters with that of groundwater taken from several wells in your study area, you can indicate which are from the same aquifers and whether there is a significant surface water component in the spring discharge. For a more detailed explanation of groundwater chemistry data interpretation, see one of the textbooks listed in the References.

Henri Schoeller, a French hydrogeologist, developed a graphical technique to examine groundwater chemistry data (Schoeller, 1962). The concentrations of the six principal chemical components – Ca, Mg, Na + K, Cl, SO_4 and HCO_3 + CO_3 – are plotted on a semi-log scale, as shown in Figure 7.10. The relationship between two chemical constituents in different water samples is shown by the slope of the straight lines connecting these constituents. Parallel lines indicate identical relationships. In this example, the chemistry of the two springs is very similar, whereas the water from the borehole is quite different, indicating that it penetrates an aquifer separate from the one supporting the springs.

Values must be converted to milli-equivalents per litre from milligrams per litre, which are the units used by the laboratory. The milli-equivalent value is calculated by dividing the concentration of each ion in milligrams by its atomic weight – which has been divided in its turn by the valency. Where the concentrations of two ions are plotted together as Na + K and HCO_3 + CO_3, calculate the milli-equivalent value of each *before* adding them together. Table 7.12 will help make these calculations easy.

You may also have to change the units to milligrams per litre before you convert the value to milli-equivalents. If you are using old records, the analyses may be expressed in parts per million (ppm); these are the same as milligrams/litre, and sometimes the units are parts per 100,000 or even

Table 7.11 *Selected chemical parameters to determine the potability of a water supply.*

Parameter	Unit	Guide level	MAC
Conductivity	$\mu S\ cm^{-1}$ at 20°C	400	1500
Hydrogen	pH units	6.5–8.5	–
Chloride	$mg\ L^{-1}$ Cl	25	400
Sulphate	$mg\ L^{-1}$ SO_4	25	250
Nitrate	$mg\ L^{-1}$ NO_3	25	50
Nitrite	$mg\ L^{-1}$ NO_2	0.10	0.50
Calcium	$mg\ L^{-1}$ Ca	100	250
Magnesium	$mg\ L^{-1}$ Mg	30	50
Sodium	$mg\ L^{-1}$ Na	20	200
Potassium	$mg\ L^{-1}$ K	10	12
Iron	$\mu g\ L^{-1}$ Fe	50	200
Manganese	$\mu g\ L^{-1}$ Mn	20	50

MAC, Maximum admissible concentration.
Based on EC Directive 80/778/EEC.

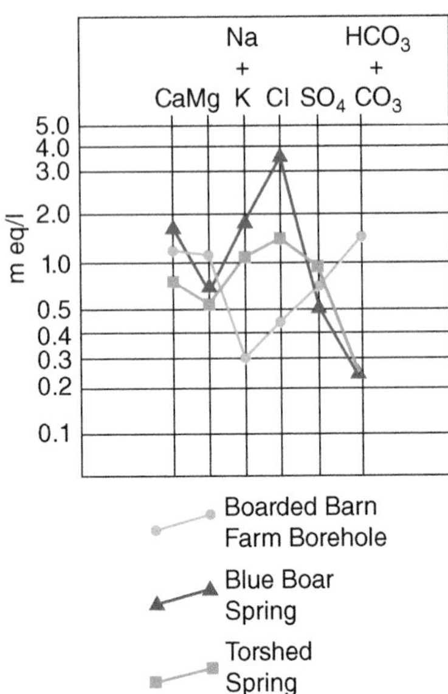

Figure 7.10 *In the Schoeller graph the relationship between two chemical constituents in different water samples is shown by the slope of the straight lines connecting these constituents. In this example, the chemistry of the two springs is very similar, whereas the water from the borehole is quite different, indicating that it penetrates an aquifer separate from the one supporting the springs.*

Table 7.12 *Conversion factors from milligrams per litre to milli-equivalents.*

Ion	Factor
Calcium (Ca)	20.04
Magnesium (Mg)	12.16
Sodium (Na)	23.00
Potassium (K)	39.10
Chloride (Cl)	35.46
Sulphate (SO_4)	48.04
Nitrate (NO_3)	62.01
Bicarbonate (HCO_3)	61.01
Carbonate (CO_3)	30.01

Milli-equivalent value is calculated by *dividing* the concentration of each ion in milligrams per litre by the factor shown.

grains per gallon. Hardness is also sometimes expressed in 'degrees'. Conversion factors for the most common are given in Table 7.13.

Another graphical method of examining groundwater chemistry data was developed by Hill (1940) and Piper (1944), and is commonly called a Piper plot, although it should be referred to as

Table 7.13 Conversion factors for various ions and analytical units.

Bicarbonate as mg L^{-1} HCO_3 = Alkalinity as mg L^{-1} $CaCO_3$ × 1.22
Hardness as Ca = Hardness as $CaCO_3$ × 0.40
1 Clark degree = 14.286 mg L^{-1} $CaCO_3$
Nitrate as mg L^{-1} N = Nitrate as mg L^{-1} NO_3 × 0.2258
Nitrite as N = Nitrite as NO_2 × 0.3043
Ammonia as mg L^{-1} N = Ammonia as mg L^{-1} NH_4 × 0.7778
Sulphate as mg L^{-1} SO_4 = Sulphur trioxide as mg L^{-1} SO_3 × 1.193
1 grain per Imperial gallon = 14.254 mg L^{-1}
1 grain per US gallon = 17.118 mg L^{-1}

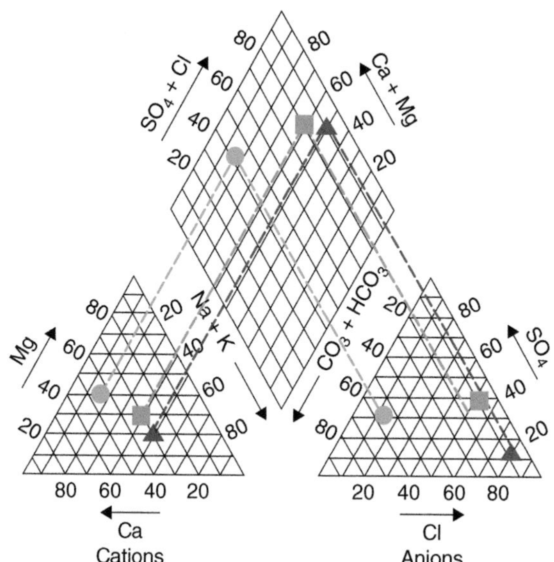

Figure 7.11 In Piper's trilinear method, concentrations of cations and anions are plotted as the percentage of the total of each when expressed in milli-equivalents. The grouping on the central diamond-shaped field shows up waters that have a similar chemical composition. Data used in this diagram are the same as those used in Figure 7.10.

the *trilinear graph* (Figure 7.11). This method examines the relative proportion of the major ions and can be used to show the effect of mixing two waters. Case History 1 shows how these techniques can distinguish between two separate groundwater bodies where only subtle differences exist.

Recharge areas can be identified with groundwater chemistry information where sufficient data are available. Plot the distribution of values for total dissolved minerals or conductivity on a map. Low values are likely to be found in the recharge areas, with groundwater flow being in the direction of increasing values. Highly mineralised groundwater is often an indication that the rate of groundwater flow is very slow, from which it follows that permeabilities are low or there are no natural discharge points. Examination of the geological structure will help you decide whether groundwater is trapped in this way.

Sometimes characteristics other than chemistry will help you distinguish between different groundwater bodies. For example, where there is thermal groundwater in the area the temperature of a spring may be a quick and easy way to identify which groundwater body you are measuring.

8
RECHARGE ESTIMATION

An important objective of most groundwater studies is to make a quantitative assessment of the groundwater resources. This can be considered in two ways: the total volume of water stored in an aquifer, or the long-term average recharge. The more significant figure in terms of groundwater resources is the long-term average recharge, and it is usual for this value to be regarded as the *available resources*. Because groundwater is a renewable resource, development will not cause a continuing depletion of aquifer storage unless the long-term average abstraction exceeds the average recharge.

It is inevitable that pumping from boreholes will cause some local lowering of groundwater levels to induce flow towards the wells. Once a new groundwater level equilibrium has been established, levels will stabilise and only fluctuate in response to annual variations in recharge, seasonal changes, and variations in abstraction. When abstraction rates exceed the average recharge, groundwater levels will gradually decline, eventually drying up shallow wells and springs and increasing the cost of pumping from deeper wells and boreholes. Eventually, groundwater may no longer be available for abstraction. Before this happens, other serious problems are likely to develop, such as a seawater intrusion in coastal areas or the up-coning of deep-seated mineralised groundwater that may cause wells to be abandoned. It follows, therefore, that resource management is essential. A knowledge of the annual average recharge is fundamental to this management process.

In Chapter 3, we saw how it is possible to estimate the volume of water stored in an aquifer by using geological information to define the total volume of water-bearing rock and the aquifer specific yield. This quantity is important as it represents a reservoir of groundwater that can act as a buffer, allowing abstraction to continue in drought years when there is little recharge.

8.1 Water Balance

Groundwater recharge to an aquifer cannot be measured directly, but only inferred from other measurements. Measurements of the other components of the hydrological cycle are used to estimate the value of the resources, using a technique called a *water balance*. It is assumed that all the water entering an aquifer system is equal to the water leaving it, plus or minus any change in groundwater storage. This can be written more fully, as in the equation given in Table 8.1. Each element of this equation is discussed in greater detail in Table 8.2, with comments on the methods of estimating each component.

The water balance method involves identifying all the inflow and outflow components that occur within an area and quantifying each one individually using field records and long-term records. Once this has been completed, the water balance equation is used to determine the groundwater recharge.

Consideration of the water balance for an aquifer can help increase an understanding of the overall groundwater flow system, even where the equation cannot be made to balance. During a study of the St Bees Sandstone aquifer, shown in Figure 4.27, the potential recharge was calculated from the effective rainfall and the area of the aquifer. The value obtained was compared with the low

Field Hydrogeology, Fifth Edition. Rick Brassington.

Table 8.1 *Water-balance equation.*

Inflows		Outflows	
rainfall + recharge from surface water + seawater intrusion + inflow from other aquifers + leakage + artificial recharge	=	abstraction + spring flow + base flow in rivers + discharge to the sea + flows to other aquifers + evapotranspiration	± change in aquifer storage

Table 8.2 *Components of the water-balance equation.*

Types of flow	Comments on methods of estimation
Inflows	
Rainfall including snow, hail and fog	This is usually the most significant recharge component. A significant proportion of rainfall is lost as evapotranspiration or runoff. Rainfall recharge is estimated using hydrologically effective rainfall values and taking account of the geology and groundwater levels.
Recharge from surface water	Where streams, rivers, lakes, or ponds have a permeable bottom or sides, water can percolate into an aquifer when the groundwater levels are lower. This recharge is estimated using Darcy's law by consideration of the geology and the difference between surface water and groundwater levels. Look for the evidence of changes in groundwater chemistry that may be caused by percolating surface waters.
Seawater intrusion	When groundwater levels in coastal aquifers are lowered by pumping, the potential exists for seawater intrusion. Apply Darcy's law (Section 3.2) to calculate inflows from information on groundwater levels and aquifer hydraulic conductivity. Look for chemical evidence to show that intrusion is taking place (Section 7.5).
Inflow from other aquifers	All aquifers that are adjacent to the study area should be examined as potential sources of recharge. Geological information, groundwater levels and chemical evidence will help decide if flow is taking place across the boundaries. Estimate inflow using Darcy's law.
Leakage	This is an artificial type of inflow caused by leakage from water-supply reservoirs, water pipes and sewers, resulting from damage or deterioration. Leakage is estimated by measuring inflows and outflows of the water supply or sewer system. In some areas, excessive irrigation of crops, parks and gardens can provide a significant quantity of extra recharge. This can only be quantified by direct observation.
Artificial recharge	In some aquifers, natural recharge is artificially supplemented by water being recharged through special lagoons or boreholes. This component of recharge can be easily quantified from direct readings. Sewage disposal may sometimes be an additional source of recharge water.
Outflows	
Abstraction	Use metered records whenever possible; otherwise estimate from the pump capacity and hours run or from the water requirements based on use. A discussion with the well operator is essential. Include other groundwater pumping, such as dewatering mines and quarries or major civil-engineering works.
Spring flows	Groundwater discharge from springs can be assessed by measuring each one separately or from stream-flow measurements, which will include the flow from a number of springs.
Baseflow in rivers	The groundwater component of surface water flows can be estimated from stream flow records (see Section 5.7).
Discharge to the sea	In coastal areas, groundwater may discharge directly into the sea. Sometimes this forms a spring-line between high and low water marks, which may be identified from observation of temperature or conductivity measurements. Apply Darcy's law to estimate the quantities involved. It can also be by flow through the seabed which is difficult to quantify.

Table 8.2 (*Continued*)

Types of flow	Comments on methods of estimation
Flows to other aquifers	Where an aquifer extends beyond the study area, groundwater may flow out of the study area, with no surface indication. Use groundwater-level information and flow-net analysis (see Sections 4.7 and 4.8) to estimate quantities. Discharges to other aquifers may occur along boundaries, and these should be estimated in a similar manner. Use geological information, groundwater-level records, and chemical information to decide if such flows can occur.
Evapotranspiration	In areas where the water table lies close to the ground surface, groundwater may be removed by plants taking water up through their roots and transpiring it out through their leaves. Generally, however, evapotranspiration removes water from the soil, thereby creating a deficit that is made good by the following precipitation. This process reduces the quantity of rainwater available for recharge. Evaporation also takes place from bare soil, with water flowing upwards under capillary forces to replace losses. In some areas, groundwater may be exposed in quarries or ponds. Here, evaporation losses will take place at the potential rate, equivalent to losses measured in an evaporation tank (see Section 5.2).
Change in aquifer storage	This is estimated from field measurements of groundwater levels and the aquifer specific yield or storage coefficient. Do not forget to take into account whether the aquifer is confined or unconfined (see Section 3.3). Sometimes the groundwater-level changes may reflect both aspects.

flows of the streams in the area (River Ehen, Ellergill Beck, Pow Beck, and Rottington Beck), which showed that much more recharge takes place than discharges through the surface water system. It was concluded that a significant part of the groundwater flow discharges into the Irish Sea. A field reconnaissance along the shore failed to find direct evidence of flow across the beach, suggesting that the main groundwater discharges are from the seabed.

8.2 Rainfall Recharge

The maximum volume of water available for recharge is the *hydrologically effective* rainfall (HER), that is, the total rainfall minus the losses from evapotranspiration. Part of this water will percolate into the ground and part will run off to surface watercourses. This portion can be assessed from stream flow gauging measurements. Evapotranspiration is the most difficult part of the hydrological cycle to quantify. Penman (1948) derived an empirical relationship between evaporation from open water and evaporation from bare soil and grass. Working in southern England, he concluded that the evaporation rate from a freshly wetted bare soil is about 90% of that from an open water surface exposed to the same weather conditions, with lesser amounts from other surfaces (see Table 8.3). Headworth (1970) showed that Penman's values are too high for the thin soils over the chalk aquifer in Hampshire, England, and more water percolates into the aquifer than Penman's model predicts.

Other workers have produced alternative methods of calculating evapotranspirational losses. Thornthwaite (1948) used an empirical approach to calculate evapotranspiration losses from grassland in the eastern USA, based on mean monthly temperature measurements. This is a widely used method, but it is really only valid in areas where it was developed and may produce significant errors if used elsewhere. Penman (1950a, 1950b) developed a more detailed approach using the concept of *soil moisture deficit*. This method allows for the fact that vegetation continues to transpire water from the soil during periods without rain, which creates a 'deficit' that is replaced by part of the percolating rainwater from later showers, thereby reducing groundwater recharge. Evaporation losses from an open water surface are measured with an evaporation tank. These values can be used to calculate field evapotranspiration losses using Penman's factors.

Penman's methods have been further developed by Grindley (1969, 1970), and more recent work has shown that recharge can still take place even during an extreme drought if the rainfall intensity

Table 8.3 *Summary of the relationship between evaporation losses from open water and those from bare soil, grassland, and chalk soils.*

Type of surface	Percentage of evaporation from open water
Open water	100
Bare soil	90
Grassland:	
Winter	60
Summer	80
Autumn/spring	70
Whole year	75
Chalk soils	25–50

is sufficiently high. These detailed aspects of recharge studies are beyond the scope of this book, and more information can be found in the textbooks in the References and Further Reading section.

In the UK, the Meteorological Office has developed the Met Office Rainfall and Evaporation Calculation System (MORECS) (Hough & Jones, 1997) that calculates values of effective rainfall for any area of the country using both long-term average and real-time assessments of rainfall, evaporation and soil moisture. The analysis covers different soil types, crops, vegetation cover, urban areas, and topography.

Factors were proposed by Rushton et al. (1988) that are applied to HER values to take account of the nature of the drift overlying the aquifers in a modelling study of a sandstone aquifer, and are reproduced in Table 8.4. The factors are based on:

information about the thickness and percentage of sand in the drift

observations about the drainage required to take away excess water following heavy rainfall

adjustments during numerical groundwater model refinement when the factors were modified to improve generated groundwater hydrographs.

Figure 8.1 shows how the factors were used to estimate recharge.

8.3 Induced Recharge

Groundwater abstraction modifies the local groundwater flow system, sometimes reversing the hydraulic gradients. This can mean that natural groundwater flow into rivers, lakes, or the sea will reduce and eventually cease. If pumping continues, water will flow out of the surface water bodies into the aquifer, depleting the flow of streams so that they will eventually dry up altogether, a situation that occurred, for example, in southern and eastern England during the period 1988–1992 where the impact of high abstraction was exacerbated by exceptionally low rainfall.

Reversal of the groundwater flow direction in coastal areas can lead to a serious deterioration in water quality as seawater moves into the aquifer. This ingress of seawater should be included in the water balance equation. Changes in groundwater levels caused by abstraction can also induce groundwater flow between aquifers.

Aquifer recharge will percolate through low-permeability layers above a confined aquifer provided that there is a downward hydraulic gradient. Figure 8.2 illustrates how such flow may be possible. In a confined aquifer the water level in wells lies within the confining layer (Figure 8.2a). The distance between the ground and this water level represents a potential excess head that will drive percolating water from the surface through the confining layer to recharge the aquifer. Such

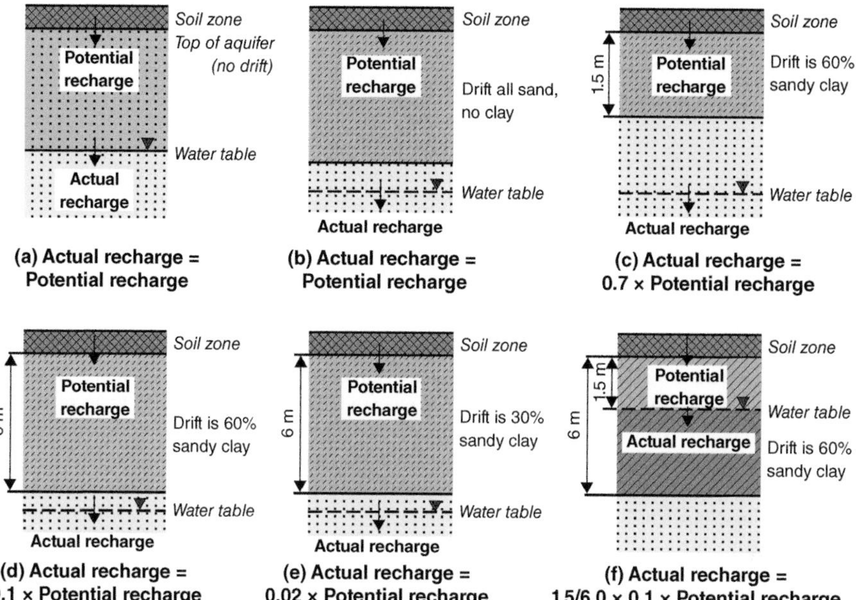

Figure 8.1 *Determination of actual recharge from potential recharge (HER) with different types of drift. (Reproduced by permission of CIWEM from Rushton et al. (1988).)*

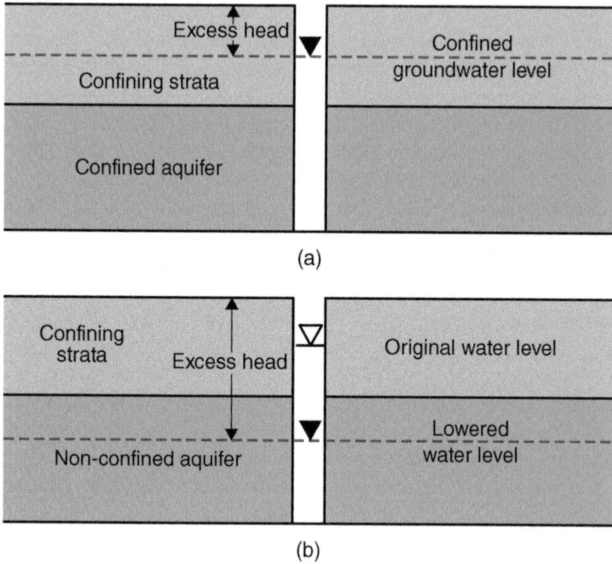

Figure 8.2 *How aquifer recharge may percolate through low-permeability layers above a confined aquifer.*

Table 8.4 *Recharge factors proposed by Rushton et al.(1988)/John Wiley & Sons.*

Clay thickness (m)	Percentage sandy clay	Factor
0	–	1
<2	50–90	0.7
>2	90–100	0.65
>2	50–90	0.1
>2	0–50	0.02

recharge is usually very small as it is limited by the low hydraulic conductivity of the confining strata and the relatively small excess head. Additionally, that part of the confining layer that is unsaturated will have a low effective hydraulic conductivity, reducing the recharge even more. However, in regional studies even such small quantities of recharge taking place over large areas of aquifer may be significant. When an aquifer becomes unconfined (Figure 8.2b) due to the groundwater level falling below the base of the confining layer, the driving head is increased to the maximum possible value. Recharge in such circumstances will be controlled by the permeability of the confining layer, and the values shown in Table 8.4 will apply.

8.4 Other Sources of Recharge

There are many potential artificial sources of water that can boost the recharge to an aquifer. For example, the excessive irrigation of crops, gardens, and public parks may recharge underlying aquifers. To assess the quantities of water from this source, calculate the potential transpiration from the plants involved and compare this with the volumes of irrigation water applied. The difference is likely to be the extra recharge.

Urban development also has a potential to modify aquifer recharge in a number of ways. Ageing water supply pipes will leak and can result in very significant volumes of water entering an aquifer. Studies have shown that such leakage may be more than 50% of the water supplied to an area and exceeds the natural recharge value. Studies have shown that even in areas with new pipes, leakage may be at least 10% of the volume supplied. Compare the total consumption with the volume supplied, or use flow meters on the distribution pipes to detect water flow at times when the water consumption is negligible. Traditionally, this is during the early hours of the morning, around 2.00 am.

Sewer pipes also deteriorate with age or may be damaged by settlement, and provide another potential source of recharge water. It is more difficult to detect sewer leakage, as there are no operational reasons for sewer flows to be measured. It may be possible to demonstrate that sewer leakage is happening using chemical evidence of relatively high concentrations of chloride, ammonium, nitrite and nitrate from excreta, phosphate and boron (found in detergent) and faecal bacteria, although these are often easily filtered out by the soil.

Urban development increases the paved areas, thereby reducing the volume of recharge. Rain falling on paved areas and roofs is usually diverted into pipes that discharge into local ditches, streams, or rivers. Where the flow is into soakaways, however, the amount of recharge may be increased. Often, storm water sewers are installed to divert the large volumes of runoff that can result during periods of intense rainfall. Holding basins may be used to contain the flood water and allow it to flow into watercourses under controlled conditions. Where the base of such flood basins consists of natural materials, there is a potential for a proportion of the water to percolate into the ground and increase the overall groundwater recharge. To make a realistic estimate, you will have to go round the area and look at the different types of ground surface and investigate the type of drainage system to assess whether recharge is feasible.

9
SPECIALIST TECHNIQUES

The earlier chapters in this book have been concerned with field measurements that you will be able to undertake yourself. Some aspects of hydrogeological investigations, however, require a specialist contractor to install piezometers, construct boreholes, undertake down-hole geophysical measurements or carry out groundwater tracing experiments. This chapter provides an introduction to these techniques so that you will be prepared to oversee the work of these specialist contractors.

9.1 Borehole and Piezometer Installation

Piezometers are small-diameter, relatively shallow tubes used to monitor groundwater levels and sometimes used for groundwater sampling (see Section 4.7), whereas boreholes are generally deeper and have a larger diameter. Both are used for gathering geological and hydrogeological information, for monitoring groundwater levels and quality, and for groundwater abstraction. The construction techniques are different and require different contractors. Make sure that you pick the right one for the job. The best way is to ask colleagues for recommendations and look for a contractor that has previously done the sort of work you need. Remember that mistakes in drilling may not be apparent until long after the contractor has left the site and been paid, and that if the borehole or piezometer is not properly installed, it may be giving very misleading information on the groundwater system that is difficult to detect and can lead to wrong conclusions being drawn.

9.1.1 Piezometers

Piezometers are installed by being driven, washed in, or drilled, with diameters typically ranging from 19 mm to about 50 mm. Driven and washed-in piezometers are normally restricted to unconsolidated materials, with solid rocks requiring drilling methods. The most common type is the standpipe piezometer, consisting of a simple vertical pipe in the ground with a perforated end. Water levels are measured using a dipper or pressure transducer and samples are taken with a bailer or an inertial pump. In some geotechnical applications, pneumatic piezometers are installed and comprise a small cylinder incorporating a diaphragm connected to the surface with a flexible tube. Any change in water level is transmitted to a pressure recording device at the surface. A similar instrument is the vibrating wire piezometer, which is in effect a pressure transducer system. Both pneumatic and vibrating wire piezometers are usually buried in the ground and therefore do not allow either check water level readings to be taken with a dipper or groundwater samples to be obtained.

Piezometer 'nests' comprise a number of standpipe piezometers installed at different depths in a deep borehole, such as the one shown in Figure 4.13. The piezometers in this installation were made using 19-mm solvent-jointed plastic pipe with a plastic permeable tip fitted to the bottom end. The tips are a perforated plastic tube about 200 mm long, sealed at one end and with a permeable plastic cylinder as an inner liner. Some versions use an unglazed permeable ceramic cylinder as the liner.

In a piezometer nest the deepest is installed first. The piezometer tube is assembled on the surface and then fed into the borehole. Being plastic, the pipe is both light and flexible and the operation can

Field Hydrogeology, Fifth Edition. Rick Brassington.

be accomplished easily by hand with two operatives. When the piezometer tip rests on the bottom of the borehole, coarse filter sand is washed into the borehole, using 50 mm diameter tremie pipe set to about 2 m or 3 m above the piezometer and a small flow of clean water, to provide about 1 m depth of sand around the tip. Once it is held in the filter sand layer, the piezometer pipe is then gently pulled straight and tied at the top to prevent movement during the remainder of the installation. A layer of bentonite pellets is installed to provide a seal on top of the sand some 0.5–1 m thick, again washed in using the tremie pipe system. The borehole is then backfilled to the position of the next piezometer, where this installation process is repeated. A sand/bentonite mix (with a 10:1 ratio by weight) is used as the backfill and is installed through the tremie pipes. Make sure that the sand is clean. Avoid using dredgings that may contain contaminants or sodium chloride if they are from the coast or tidal estuary. It is possible to install up to six 19 mm diameter piezometers in a 200 mm diameter borehole. To test whether the seals are effective, measure the water level in each piezometer. If the seals are working, the levels should all be different!

9.1.2 Dynamic probing

Dynamic probing is a technique used to obtain information on the strength of soils by driving a standard-sized probe into the ground by hitting it with a standard-sized slide hammer that falls from a standard height onto it. The end of the probe is pointed to help it penetrate the ground. The number of blows that it takes to drive increments of 10 cm provides a measure of the soil strength. The value of the method to hydrogeologists, however, is that the hole created by the probe can be used to insert a standpipe piezometer. Sometimes the probe is replaced with a steel pipe that has a specially strengthened piezometer tip fitted with a pointed drive head. The method can only achieve depths of about 6 m, and usually much less. It compacts the soil and geological materials that the probe penetrates, which is likely to reduce its permeability. Nevertheless, it can provide a quick and cheap method of installing piezometers for groundwater monitoring.

9.1.3 Jetting

Pipes with diameter up to 50 mm can be washed into unconsolidated materials up to coarse gravel or small cobbles in size. The method is commonly used to install wellpoint systems used for dewatering on construction sites where the water table needs to be lowered. Typically, 6 m of pipe is screwed together and a flexible hose is connected to one end. A small starter hole is dug and the end of the pipe stood vertically in it. The pipe and hose are very heavy when filled with water and it is usual for them to be supported on the arm of a back-acting excavator. A water supply of about 20 L s^{-1} is commonly used and is fed down the hose at high pressure. A jet of water squirts from the open end of the pipe and turns the granular soil around its base into slurry so that the pipe can be pushed into the ground. The soil at the greater depth is made into slurry, allowing the pipe to move down. The method is rarely used to construct piezometers except where they are being used round wellpoint systems.

9.1.4 Shell and auger

Shell and augur is the most common drilling method used in geotechnical site investigations for civil engineering construction. Shell and auger (or light percussion drilling) comprises a collapsible ‘A’ frame, with a pulley at its top. A wire rope runs from a winch drum over the pulley and is used to raise and lower a series of weighted tools into the borehole. The winch is powered by a diesel engine and is operated using a foot-operated friction brake (see Figure 9.1). Such rigs are very light, easily towed, and simple to erect. Where access is restricted, the rig can be dismantled and rebuilt where the hole is to be drilled.

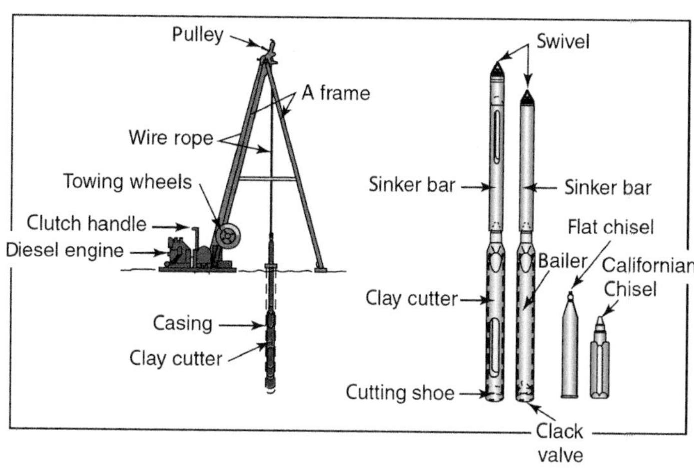

Figure 9.1 *The main features of a shell and auger rig. Tools shown are a clay cutter, bailer, and chisels.*

The tools are simple and made out of steel. A 'clay cutter' is used to drill in clays and materials that contain enough clay to make them stick together. It is a heavy steel tube with a sharpened end. Its weight drives it into the bottom of the borehole so that clay becomes wedged inside it. The driller pushes the contents out of the clay cutter using a metal bar that is inserted through the open slots in the side of the tube.

The tool used to drill through sands or gravels is a 'shell' that is also a steel tube, although in this case it has a flap on the bottom end that forms a simple non-return valve. Water is poured into the borehole, and the shell is surged up and down every second or so over a distance of about 300 mm. The upward movement sucks water into the bottom of the hole, loosening the material and forming a suspension or slurry. The non-return valve opens on the down stroke, allowing the slurry to flow into the shell. The valve remains closed when the shell is pulled out so that the cuttings can be emptied out at the surface. Geological information is obtained by inspecting the cuttings from the shell and the material removed from the clay cutter. Problems in the accuracy of the driller's log occur when material falls into the hole from above. A chisel is used to smash through hard objects such as boulders and to obtain some samples from the rockhead so that the rock type can be identified.

The borehole is supported by steel tubes or casing that is driven down the hole every time its base is extended by about 200–300 mm. When boring has finished, the casing is removed using jacks and the borehole backfilled with suitable material. Good practice requires this to be impermeable grout, but all too often no seal is provided, thereby creating a potential flow path for water along the borehole. This may cause pollution to enter an aquifer or, conversely, may allow artesian groundwater to flow out at the surface, resulting in localised flooding. When a standpipe piezometer is to be installed, the pipe is assembled at the surface or screwed together as it is lowered into the borehole. Filter sand is placed round the tip with a bentonite seal, as described in Section 9.1.1. The borehole is then infilled and the casing removed in stages to avoid any collapse that would ruin the piezometer installation.

9.1.5 Percussion drilling

Cable tool or percussion drilling is the oldest method of water well construction. It is very similar to the shell and auger method described above, although the size and weight of the equipment are very much greater, and the two systems should not be confused. The method is still in use, although it has been largely replaced by hydraulically driven, top-drive rotary techniques.

Heavy steel tools are used that are screwed together to form the tool 'string'. They are suspended on steel wire rope that has a left-hand twist which gives a slight rotation to the tool-string and keeps the right-handed tool joints tight. The rope passes over a pulley at the top of the mast and round the ends of a pivoted beam (spudding beam), and is wound on a heavy-duty winch. The beam moves up and down to give a controlled cutting movement to the end of the tool suspended at the bottom of the borehole. The slight rotation of the tool means that it never hits the same place twice and so cuts a circular hole. The cuttings are removed as slurry using a bailer (or shell) that is run down into the borehole on a separate cable called the 'sand-line'.

Different shaped bits are used for hard and soft geological formations. Soft ground chisels have a sharp edge, generous clearance angles and large waterways to allow rapid movement through viscous slurry. Chisels for hard formations have a blunt cutting edge and a large cross-section area to give strength, and have 'flutes' or waterways on the sides. The string of tools also includes a drill stem comprising a long steel bar used to add weight to the drill bit to help cut a vertical borehole. Drilling jars are fitted above the stem and consist of a pair of steel bars linked together in the same way as a chain. If the tools become stuck, the line is slackened, which allows the links to close together. Pulling sharply up on the line gives an upward blow to the string of tools and is more successful in freeing the tools than a steady pull on the rope, which often breaks it. Figure 9.2 shows the general arrangement for a cable-tool drilling rig and the main tools that are used.

9.1.6 Rotary drilling

In rotary drilling, the borehole is excavated by a rotating drill bit fitted to the end of a drill pipe 'string'. The top rod is screwed into the drive head that incorporates a hydraulic motor powered by a diesel engine. This rotates the drill string so that the bit rotates at speeds of 30–300 rpm, depending on the diameter and strata. A roller rock bit is generally used and has three hard-steel toothed cutters that are free to run on bearings. Drilling fluid is pumped at high pressure down the drill pipe, where it emerges through ports in the drill bit, flowing over the rollers to lubricate them, clean the teeth and carry away the cuttings. The fluid returns to the surface up the borehole, carrying cuttings that are settled out in a pit or tanks. Depending on the fluid, it may be reused down the borehole once the cuttings have been removed. Figure 9.3 shows the main features of a rotary drilling rig.

As drilling proceeds, the drive head travels down until the swivel unit reaches the top of the clamp. Drilling is stopped and the drill string is raised a short distance to move the bit off the bottom of the borehole. Fluid circulation is continued for a short time to carry the most recent cuttings up the borehole and clear the bit. The pump is then stopped and the top length of drill pipe is held by a clamp that fits into flats on the side of the pipe. It is then unscrewed from the drive head and the drill string is held by the clamp resting on the top of the casing. The head is then raised and tipped to the side to allow another drill pipe to be screwed onto it. This is then lowered so that the bottom of the new length of pipe fits into the top of the drill string. The pipe is screwed together and the clamp removed so that the drill string can be lowered until the bit touches the bottom and drilling is restarted. This procedure is repeated until the final depth is reached.

Compressed air is the most usual drilling fluid for small-diameter rotary boreholes. However, it can only be used to depths of about 150 m below the water table and diameters up to 300 mm, because the water pressure counteracts the air pressure, gradually slowing the circulation rate. For depths greater than 150 m a larger compressor is required, although other fluids are used for drilling to greater depths because the large-capacity compressors needed are prohibitively expensive.

Foam drilling is a variation of air flush where a foaming agent is mixed with water and injected into the drill stem with compressed air. The foam is formed as the mixture travels down the drill pipe, reappearing at the surface with its load of cuttings and breaking down in 30–45 minutes. Rapid

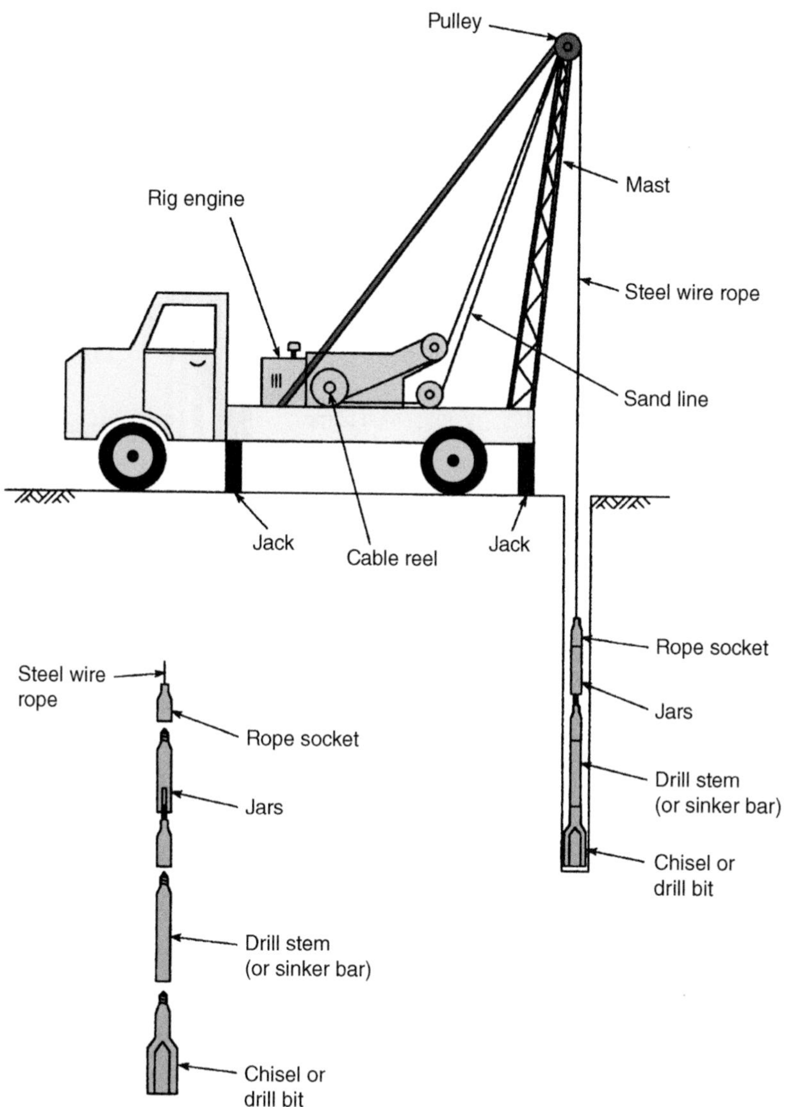

Figure 9.2 *A lorry-mounted percussion rig and the common tools that are used.*

drilling rates are usually achieved and a cleaner hole is produced than with most other methods. The same depth limitations apply as to normal air flush.

A number of proprietary muds are available for water well drilling that are designed to have physical properties to improve the drilling rate, move the cuttings efficiently, support the borehole wall and seal it to minimise drilling fluid loss. These muds are non-toxic and made from types of vegetable or bacterial gum, and need to be removed from boreholes by simple chemical treatment, or some break down after a short period and can be removed by pumping.

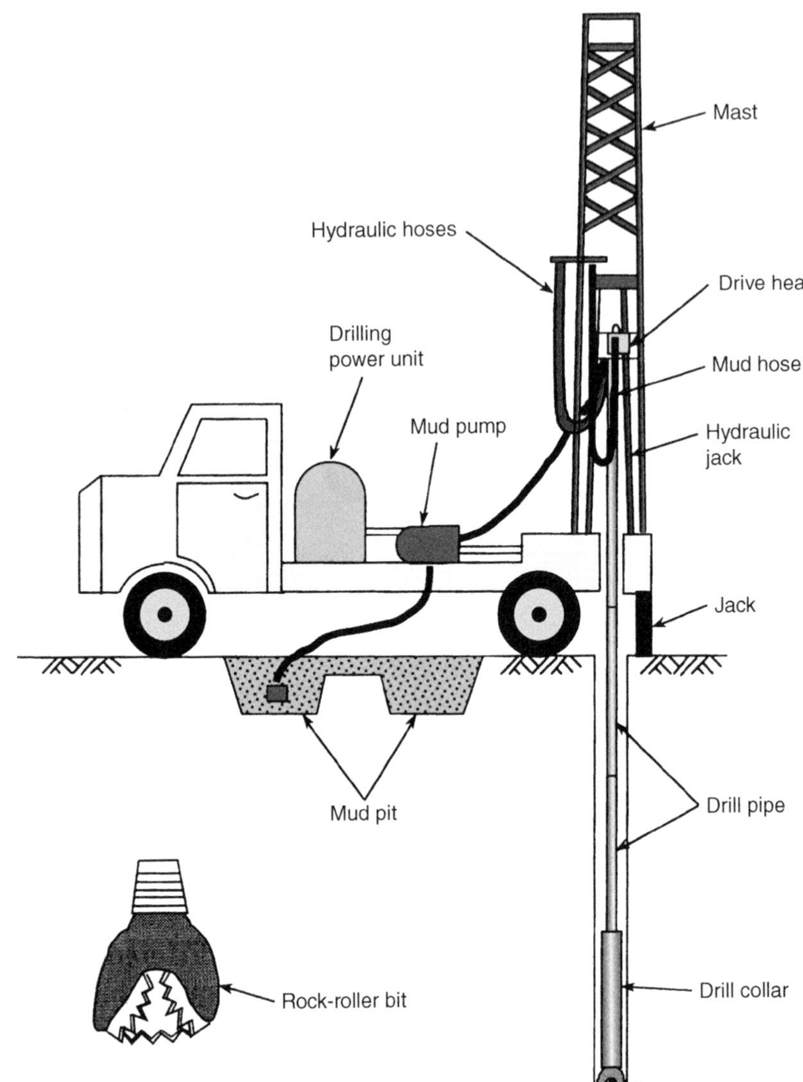

Figure 9.3 *The main features of a lorry-mounted rotary drilling rig.*

Any coating of cuttings that builds up on the borehole face must be removed or it will reduce the yield of the borehole. This process is called 'development' and involves scraping the well face or inducing high velocities by pumping or using a form of plunger. Chemical treatment is also used to help break down this coating.

Where core samples are required, a core-barrel is used instead of a rock bit. The core-barrel comprises a steel cylinder of varying length between 1 and 6 m. The rock is cut by a 'crown' fitted on the bottom end of the barrel, which either has teeth that do the cutting or is encrusted with hardened steel particles or industrial diamonds. The type of crown is selected according to the hardness of the rock being drilled. Progress is slow, as the entire drill string must be removed from the borehole to recover the core each time the barrel is full.

9.1.7 Rotary drilling – reverse circulation

Reverse circulation techniques are used to drill large-diameter boreholes. Water is the main drilling fluid used and is allowed to flow down the borehole outside the drill pipe and is returned to the surface up the drill rods, hence the name. This is achieved by the borehole being drilled through a small pool, with the water level maintained by a constant water feed. Drilling water flows down the borehole under gravity and is pumped up through the drill pipe using a suction pump. It is discharged into a settling lagoon and the settled water flows back into the pool and recirculates down the borehole. Water is lost to the aquifer as the natural groundwater level is usually below the water level in the pool. It is important to ensure that only clean water is used for this drilling to reduce clogging the well face. The equipment is broadly similar to that used in direct circulation drilling but considerably larger and heavier. The waterway through the tools and drill pipe is typically 150 mm or more. The minimum practical drilling diameter is approximately 400 mm, with the method being capable of constructing boreholes with a diameter in excess of 1.8 m.

9.1.8 Down-hole hammer

The down-hole hammer method is a modification of normal circulation air drilling and has significant advantages of rapid penetration and coping with hard rocks. The drill bit is an air-actuated single-piston hammer fitted with tungsten carbide cutting teeth. The hammer works on the same principle as the familiar road-drill and is capable of very fast drilling rates. The complete assembly is rotated at 20–50 rpm to change the position of the cutting teeth and produce a circular hole. The bit strikes the bottom of the hole at 250–1000 blows per minute and cuttings are removed by the air circulation. Diameters can be drilled up to 750 mm. The same limitations in depth apply as with more conventional air flush drilling.

9.2 Down-Hole Geophysics

Surface geophysics provides a number of field methods to supplement your geological information, such as those described in *Basic Geological Mapping* (5th edition 2011) by Lisle et al. Down-hole geophysics involves lowering special electronic instruments down a borehole to measure various physical properties of aquifer rocks and the fluids they contain. In hydrogeological studies these measurements are used to help identify the aquifer rocks, to increase the information on groundwater chemistry and aquifer behaviour, and to assess the construction and performance of water wells. The equipment is normally truck mounted, with a winch for lowering the equipment and computers to control the measurements and record the data, as can be seen in Figure 9.4, including a variety of different tools. The geophysical instrument (usually termed a 'sonde') has been lowered down the borehole on a multi-core cable that passes over the pulley set up over the borehole top. The operator is controlling the winch and the instrument through a laptop computer. Data from the sonde is recorded on the hard drive and can be processed to produce logs similar to those in Box Figure 4.1. The methods and application to hydrogeological studies are described by Beesley (1986). Most down-hole geophysics in groundwater chemistry studies are carried out by a few specialist contractors, although some university departments and a few large consultancy firms may have their own equipment.

Sondes have the electronics and sensors fitted into a steel tube some 50 mm in diameter and 1 or 2 m long. The sonde is lowered to the bottom of the borehole and then hauled out again on multi-core cable at controlled speeds. A continuous digital recording is made as the sonde travels up the

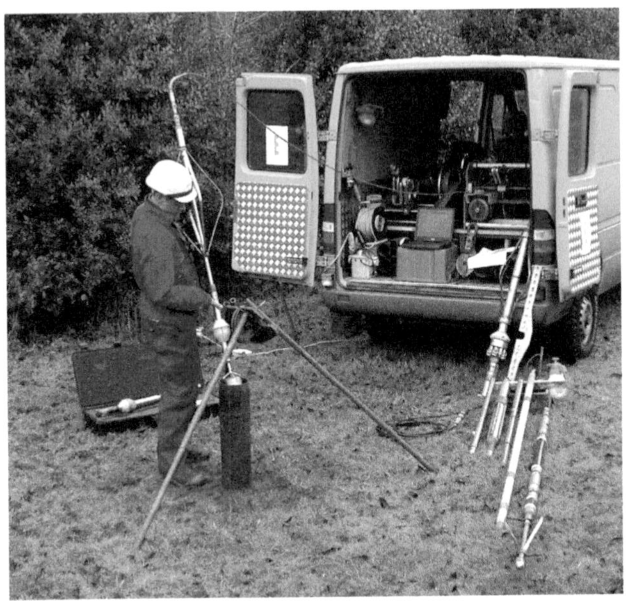

Figure 9.4 *Down-hole geophysical measurements being taken. (Courtesy of European Geophysical Services Limited.).*

borehole, relating each measurement to its depth. The data are usually presented graphically in reports, with each parameter shown as a vertical profile on the same vertical scale. The scale for each measurement is shown at the top of the graphs and the depth along the side. Comments taken from the CCTV inspection are also shown as a depth profile. Case History 4 gives an example of how these data are both presented and interpreted.

Case History 4 – The Use of Borehole Geophysics

A water-supply borehole was evaluated using a CCTV survey and the down-hole geophysical logs shown as graphs in Box Figure 4.1. The borehole penetrates two sandstone aquifers separated by shales and overlain by a thick sequence of glacial drift, as shown in Box Table 4.1. A sand and gravel deposit lies directly on the upper sandstone rockhead and is thought to be a key part of the aquifer system supplying the yield.

The caliper log showed the casing to have an internal diameter of some 305 mm and to extend to a depth of 56.3 m (also seen on the CCTV). The base of the casing is a fixed point and can be used to compare the geophysics data with the driller's log (Box Table 4.1). The log shows the base of the casing at 57.9 m depth, some 1.6 m lower than that found by the geophysics. Such anomalies are common and caused by the use of different datum points. By relating all measurements to the base of the casing it is possible to confidently compare the geophysical results with the driller's log.

The caliper log shows that the diameter of the lower, unlined section of the borehole is about 320 mm to a depth of 65.75 m, where it decreases to some 300 mm to the bottom of the logged section. The caliper log indicates much fracturing in the section 56.3–65.5 m, with occasional fractures below that depth.

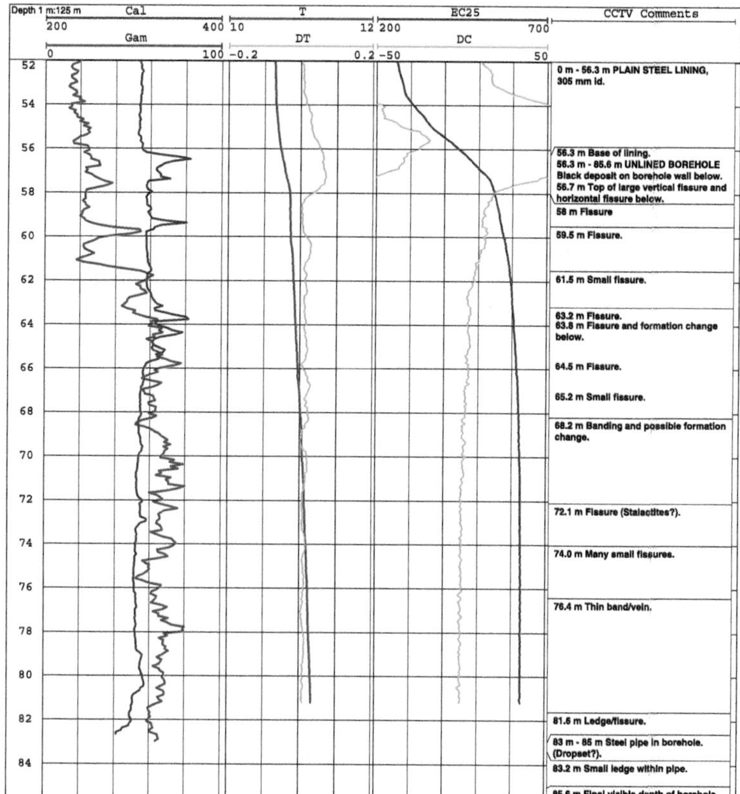

Box Figure 4.1 *Vertical profiles of caliper (Cal), natural gamma (Gam), fluid temperature (T), differential temperature (DT), fluid conductivity (EC25), differential conductivity (DC) and comments on features seen during CCTV inspection.*

Box Table 4.1 *Borehole geological and construction details.*

Formation	Description	Thickness (m)	Depth (m)
Made ground	–	1.5	1.5
Drift – confining layer	Silty clay	28.1	29.6
	Silt	10.6	40.2
	Gravel	7.9	48.1
	Clay	0.2	48.3
Drift (part of upper aquifer)	Gravel and sand	8.3	56.6
Upper sandstone aquifer	Sandstone	9.4*	66.0
Aquitard	Blue shale	4.4	70.4
	Shale with sandstone bands	11.9	82.3
Lower sandstone aquifer	Sandstone	10.0*	92.3

* Borehole does not penetrate full thickness of formation. Rest water level 4.5 m depth. Casing: 305-mm tubes to 57.9 m depth.

The CCTV inspection showed that the casing was heavily blistered and encrusted. Particles of rust were visible in suspension throughout most of the borehole water column and very high concentrations of the total iron were found in the chemical analyses of a water sample. However, the filtered sample has an exceedingly low iron content, indicating that most of the iron in the total iron analysis is from the borehole casing.

The base of the borehole was found to be at a depth of some 85.6 m, compared with the recorded total drilled depth of 91.3 m below the same datum. Allowing for the differences in the datum points, this means that the bottom 4.1 m of the borehole is filled with debris that appears to be a mixture of rock from the borehole sides and iron particles from the casing.

The CCTV showed fractures in the rock at various depths. Significant fissuring at 56.7 m is shown to be vertical and about 2 cm wide, and is one of several that occur in the sandstone section immediately below the casing. These fractures are probably joints in the upper sandstone and likely to make a significant contribution to the borehole yield.

The gamma log shows a significant change in the nature of the geological formation at 63.8 m, which is interpreted as the base of the upper sandstone aquifer and is similar to the value on the driller's log of 64.4 m depth when adjusted for the difference in datum. The slightly greater depth of 0.6 m shown on the driller's log was probably caused by fragments falling into the borehole from the fractured sandstone. Such anomalies between drillers' records and geophysical logs are not unusual.

Below 63.8 m the formation has gamma values in the range 60–80 units, with a few small-scale variations indicating the presence of alternating thin sandstone and mudstone (or probably siltstone) bands, as described in the driller's log, which records the top of the deeper sandstone aquifer at some 80.7 m depth below the new datum. The gamma log extends to about 83 m depth but does not show the top of this lower sandstone.

The temperature log showed a gradual increase from about 10.6°C at the base of the casing to about 11.2°C at 81 m depth. This represents a gradient of around 2°C/100 m and is typical for the regional geothermal gradient. The differential temperature log shows activity at around 57 m, 61 m and 67 m, which is interpreted as being inflows at the deeper points and an outflow at around 57 m.

The conductivity log shows a marked change, from 270 $\mu S\ cm^{-1}$ inside the casing to some 630 $\mu S\ cm^{-1}$ at 67 m, below which there is little change. This implies an upward flow from the deeper sandstone into the upper aquifer. The differential conductivity showed major flow activity just below the casing, which is consistent with this interpretation.

The geophysical tools that are typically used in hydrogeological work are listed in Table 9.1. Temperature and conductivity profiles for the groundwater column in the borehole are normally presented both as absolute readings and as a differential comparing the value measured at each point with that measured at a set distance above it (usually 1 m). These data are useful in helping to understand the flow conditions in a borehole and are often used in conjunction with a caliper log to identify inflows and outflows in the borehole. Water flow in the borehole is measured directly with either an instrument fitted with impellers or a heat-pulse flow meter. The latter works by rapidly heating a small volume of water and detecting the direction of flow with thermistors (electronic thermometers) set above and below the heat source. Conductivity readings also indicate changes in the chemistry of the water column in the borehole.

Caliper and natural gamma logs are also commonly used in hydrogeological studies. A caliper sonde is fitted with three or four spring-loaded arms, hinged rather like the spokes on an umbrella. It is lowered to the bottom of the borehole; the arms are then opened and the instrument pulled back to the surface so that the end of each arm touches the borehole sides. The log shows fissures and changes in diameter and can even pick up the joints in the borehole casing. A gamma log records the natural gamma emissions produced by the radioactive decay of potassium that is abundant in clay minerals. Hence the gamma log produces an indirect measure of the clay content of the formation.

Table 9.1 Geophysical tools used in hydrogeology.

Conductivity	Measured as a continuous profile of the borehole water column. Data are usually presented at a standard temperature of 20°C (or 25°C) as an absolute reading for a particular depth. Values are given in micro-Siemens per centimetre ($\mu S\ cm^{-1}$)
Differential conductivity	Measured as the difference between two points 1 m apart to accentuate differences. Useful to identify inflow or outflow points and flow along the borehole.
Temperature	Measured as a continuous profile temperature of the borehole water column. Data are usually presented in degrees Centigrade (°C) as an absolute reading for a particular depth.
Differential temperature	Measured as the difference in temperature between two points 1 m apart. Used to identify inflows and outflows and flow along the borehole.
Flow logs	Measurement of water flow within the borehole. Low velocities are difficult to detect and coloured plastic streamers attached to the CCTV camera may be more effective.
Natural gamma	Provides an indirect measure of mudstones and clays penetrated by the borehole using the natural gamma radiation emitted from potassium, thorium, and uranium that are common in clay minerals.
Caliper	Measures variations in borehole diameter. Some versions present two sets of data set at 90° to each other. Useful in examining construction details of a borehole and in identifying fractures.
CCTV	Presented as a DVD (or sometimes as video) colour recording of a continuous survey of the borehole sides in two aspects: the first looking downwards along the axis of the borehole, and the second, a survey looking sideways. Useful in examining construction details of a borehole and in identifying fractures.

The interpretation of how a suite of logs is used to examine a water well and assess its condition is illustrated by Case History 4.

9.3 Using Artificial Tracers

Tracing involves adding a substance to water in either solution or suspension at the point where it disappears below the ground, and then monitoring all possible outflow points for signs of its reappearance. This identifies links between sinkholes where water disappears underground and springs or other resurgences, and allows travel times to be calculated. Besides an important role in karst aquifer studies, the technique is also used to help understand flow mechanisms, to quantify resources and to locate the source of a pollutant. In simple terms, tracers can show that water entering an aquifer at point A arrives at point B (and possibly points C, D, and E as well), with a given travel time. Do not fall into the common trap that if no tracer is recovered there is no connection between the injection and monitoring points. It is more likely to mean that you chose the wrong tracer and it has become trapped in the flow system, or you have not been monitoring long enough and all the tracer will flow out after you have packed up and gone home.

An ideal tracer should be detectable in minute concentrations, should not occur naturally in the tested waters, and should neither react chemically with the geological materials present nor be held in them by absorption. It should be safe in terms of health and the aquatic environment, and should be cheap and readily available. Unfortunately, there is no single substance that is suitable as a tracer in all situations and so many different ones have been used, as described below. Those most commonly used in groundwater studies are fluorescent dyes and optical brighteners from the textile industry, and bacteriophages, which are viruses parasitic on faecal bacteria. This small group of tracer materials has the advantages of being effective in most situations and being acceptable in terms of their potential environmental impacts and impacts on human health. However, a large number of other materials can be used as tracers and may be more effective in certain situations. The selection and use of tracers is a specialist area of hydrogeology, and a detailed guide is provided by Ward et al., (1998) and White (1986).

9.3.1 Dyes

A distinctive colour is given to the water using a chemical dye that is visible in low concentrations. Detection is either by direct observation or by concentrating the tracer onto hanks of cotton or absorption on a ball of activated charcoal. Some of the dyes used are fluorescent and can be detected using a fluorimeter. Either water samples are taken at regular intervals for laboratory testing or charcoal balls are used as collectors. These collectors are about the size of a golf ball, with the activated charcoal wrapped in stocking-type material. This method has the advantage of concentrating the sample over a long period and is good in situations where the only outcome required is to establish a link. The presence of the dye is detected using a fluorimeter in the laboratory. Commonly used dyes are rhodamine WT and fluorescein (uranine) and optical brighteners such as photine CU and Tinopal ABP. An example of the use of these tracers is provided in Case History 5.

Case History 5 – The Use of Groundwater Tracers

A tracer experiment was undertaken to identify connections between a major quarry in the Carboniferous Limestone in northern England and local springs and streams. The limestone is well jointed and shows a number of karst features, including swallow holes and enlarged joints exposed both on the limestone pavement and in the quarry workings. The quarry is about 0.75 km^2 in area and is dug into the southern slopes of a hill, with a river system to the south. The limestone forms the hill and extends to the south, where it is overlain by clayey alluvium, with most of the springs occurring along the edge of this alluvial deposit, as shown in Box Figure 5.1.

Surface runoff from above the quarry and rain falling directly onto it accumulates in several places on the quarry floor, which is more than 15 m above the local water table. The water is discharged into the ground at three locations, comprising a purpose-made soakaway on the southern edge of the site (at some 55 mOD) and two solution-enlarged joint fractures exposed in the quarry floor at 75 mOD and 95 mOD. The Environment Agency required the quarry operator to demonstrate where the water that flows into the aquifer at each point emerges in the stream system, so that the discharges could be given the appropriate authorisations.

Springs and groundwater discharges were identified over an area some 2.5 km from the quarry, using information from the 1:25,000 scale topographical map and verified by a walk-over survey. Local residents were asked for information on the sustainability of the springs and to point out any not already identified from the map. These springs lie at elevations in the range 28–35 mOD.

The experiment was started on 1 November, with separate tracer chemicals being injected into the three discharge points. Rhodamine WT was used at the soakaway (A in Box Table 5.1), fluorescein at the fracture on the western side of the quarry (fracture B) and Tinopal ABP at the fissure system in the northern corner (fracture C). The tracers were detected using gas-activated carbon receptors placed in the 18 springs thought to be potential emergence points. The results of the tracing exercise are summarised in Box Table 5.1 with the 'first show' dates for the three tracer chemicals at the seven locations where they were detected. The injection and monitoring points are shown on the map in Box Figure 5.1.

The activated carbon receptors were not recovered daily due to limited resources. Receptors had been placed at each site about a week before the tracers were injected to identify any background material that could be mistaken for the tracers. The receptors were changed when the

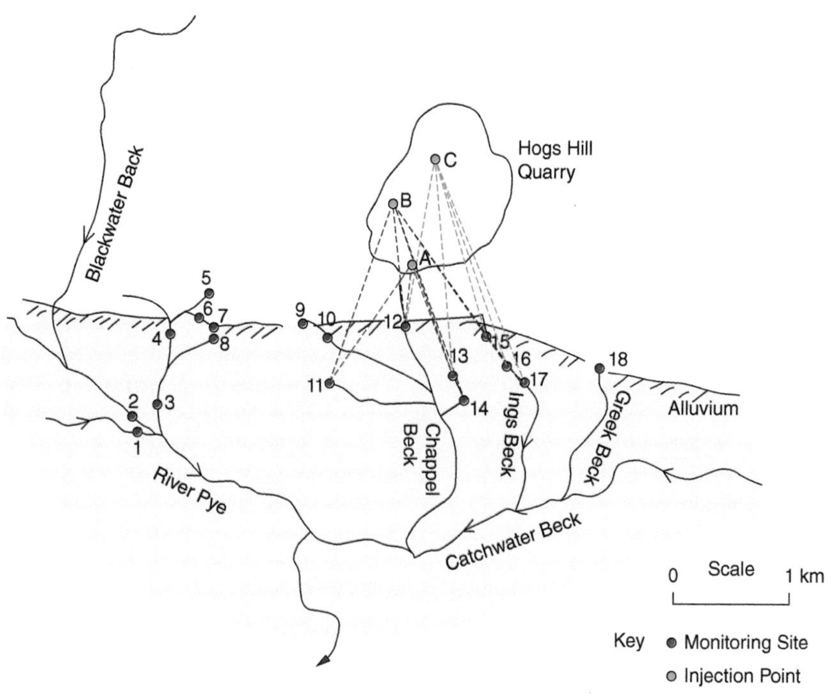

Box Figure 5.1 *Location of sites discussed in the text, including the three injection points and 18 monitoring points in relation to the geology and drainage system. Monitoring points are mainly springs at the alluvium/limestone contact, with a few that act as controls.*

Box Table 5.1 *Summary of tracer experiment at Hogs Hill Quarry, northern England.*

		First show dates		
Site	**Description**	**Soakaway (A) Rhodamine WT**	**Fracture B Fluorescein**	**Fracture C Tinopal ABP**
1	River Pye – non-limestone used as control	nd	nd	nd
2	Blackwater Beck	nd	nd	nd
3	Spring west of Underhill Farm	nd	nd	nd
4	Underhill Beck	nd	nd	nd
5	Ephemeral spring	nd	nd	nd
6	Spring in stone culvert reputed to be permanent	nd	nd	nd
7	Spring at head of drainage channel upstream of site 2	nd	nd	nd
8	Spring in stone culvert	nd	nd	nd
9	Spring at head of tributary	nd	nd	nd
10	Spring flowing from stone-lined culvert	nd	nd	nd

(Continued)

Box Table 5.1 (Continued)

Site	Description	First show dates Soakaway (A) Rhodamine WT	Fracture B Fluorescein	Fracture C Tinopal ABP
11	Spring at end of track	15 November	15 November	nd
12	Springs at head of Chapel Beck	2 November	2 November	2 November
13	Spring in field 100 m south of church	2 November	2 November	2 November
14	Stream 200 m south of site 13	nd	4 November	nd
15	Main feeder to Mill Pond	nd	2 November	2 November
16	Main outlet to Mill Pond	nd	2 November	2 November
17	Culverted spring rises at Ings Farm	nd	9 November	9 November
18	Spring at head of Green Beck	nd	nd	nd

Injection on 1 November.
nd, None detected.

injections were made and then replaced on the following day and on days 3, 8, 14, and 30 after the injection.

The experiment showed that fracture B is connected to the widest network of groundwater conduits, with tracer being seen at seven springs. Fracture C was shown to be connected to five of the springs and was not seen at the two receptor sites that have the longest travel times for the fracture B tracer. The soakaway (A) is connected to only three springs, suggesting that it intercepts a small fracture network. The most rapid flows were from all three injection points to the springs at sites 12 and 13, and from the two natural fractures (B and C) to the mill pond at sites 5 and 16, with travel times less than 24 hours. The slowest flows were from the soakaway (A) and fracture B to site 11, which took between 8 and 14 days, presumably travelling through a narrow fracture system or one with a convoluted flow path. The injection chemical from fracture B was detected as a trace after 30 days at sites 12, 13, 15, 16, and 17, and that from the soakaway at site 12, suggesting that the tracer pulses had almost tailed off by that time.

The experiment provided information on the links between the discharge points and the locations of each emergence, together with an indication of variations in the geometry and therefore the flow resistance of the fracture network.

9.3.2 Solutes

A concentrated solution of inorganic salt is added to the water at the sinkhole. The usual chemicals are soluble chloride and sulphate salts, and sugars, with the most common being sodium chloride. The tracer is detected by frequent sampling followed by chemical analysis. Electrical conductivity changes significantly with increases in dissolved minerals, and is used as an indirect measure of changes in concentration. The measurements are made in the laboratory or on site using a portable probe. A continuous record can be obtained by connecting the probe to a data logger that allows travel time to be calculated in detail.

Leaking mains water is a special case as it may have its own built-in tracer. Mains water is treated using chlorine gas (or compounds such as sodium hypochlorite (NaClO) that release chlorine gas) to ensure that any bacteria it contains are killed. If the water contains any organic material, such as humic acids, a reaction will occur, forming a series of compounds known collectively as trihalomethanes (THMs). The THM chemicals are simple single-carbon compounds with the general formula CHX_3, where X is any halogen (i.e. chlorine, bromine, fluorine or iodine) or any combination of these. The most frequently detected THM compounds are chloroform (trichloromethane), bro-modichloromethane, dibromochloromethane and bromoform (tribromomethane). THMs can only be detected by laboratory analysis, which typically includes a total THM measurement and tests for the representative THM compounds. Detection limits are usually in the range of 0.2–0.4 $\mu g\ L^{-1}$. Tests for any free chlorine present are usually carried out on site using a kit comprising a small test tube that is filled with a water sample to which a tablet of indicator chemical is added. The colour of the water is compared to a standard chart to determine the presence and concentration of the free chlorine.

THMs are completely artificial and do not occur in nature. If any are detected in water, there are only two possible origins: mains water treatment and leaks from the chemical plants that make them. Their presence usually provides unambiguous proof that a water main is leaking and shows a link between the local water distribution network and the site where the water sample was taken. The use of these chemicals as tracers is restricted, as both the THMs and chlorine gas are volatile and gradually dissipate from the water. Consequently, the method may not work at large distances from the leaking water main and usually works in a few tens of metres from the leak site. The loss of the chemicals from the water is time related, and so the distance depends on the permeability of the ground that controls the flow rates from the leak.

I was asked recently if a flooded basement was caused by groundwater. There were no obvious signs in the local area and so I suggested that a sample be taken and analysed for pH, electrical conductivity, calcium, magnesium, sodium, potassium, chloride, sulphate and alkalinity that would characterise the water; trihalomethanes (total), 1,2-dichloroethane, 1,1,1-trichloroethane, tetrachloromethane, benzene, trichloromethane, trichloroethene, dichlorobromomethane, dichlorobromomethane, tetrachloroethene, tribromomethane, and chlorine to test for the presence of treated mains water; and total coliforms, faecal coliforms, enterococci, escherichia coli, faecal streptococci, TVC (1 Day @ 37 °C), TVC (3 Days @ 22 °C), TVC (2 Days @ 37 °C), aeromonas hydrophilia, clostridium perfringens and giardia to test for the presence of sewage.

The results came back showing the water as being water type $Ca\text{-}Mg\text{-}HCO_3\text{-}SO_4$ as can be seen in Figure 9.5. The tests for treated mains water showed chloroform of 1.19 $\mu g\ L^{-1}$, bromodichloromethane of 1.52 $\mu g\ L^{-1}$, and dibromochloromethane of 0.83 $\mu g\ L^{-1}$ with a total TMS result of 3.55 $\mu g\ L^{-1}$. The presence of sewage was shown by low concentrations of all the determinands set out above. The conclusion drawn was the water was derived from a leaking water main that had raised the local water table to above the elevation of the basement floor. The water flowing through the ground had picked up the sewage effluent from a local septic tank effluent.

9.3.3 Particles

Many early tracing experiments were conducted using a suspension of fine particulate matter added to the water and detected either by visual observation or by capture on nets. The most commonly used material was spores from *Lycopodium* (a species of club-moss) that are typically 30 μm in size. Where tracers are to be added to several sinkholes, the spores are dyed to identify their origin. This technique allows up to six sinkholes to be tested at once. Other mechanical tracers, including small beads of polypropylene, are also sometimes used, although it has been reported that spores are the only successful tracers of this type.

9.3.4 Microbes

Microbes are sometimes used as small particles to trace groundwater movement, and fall under two headings, bacteriophages and bacteria. The bacteriophage (often just called phage) used is a culture

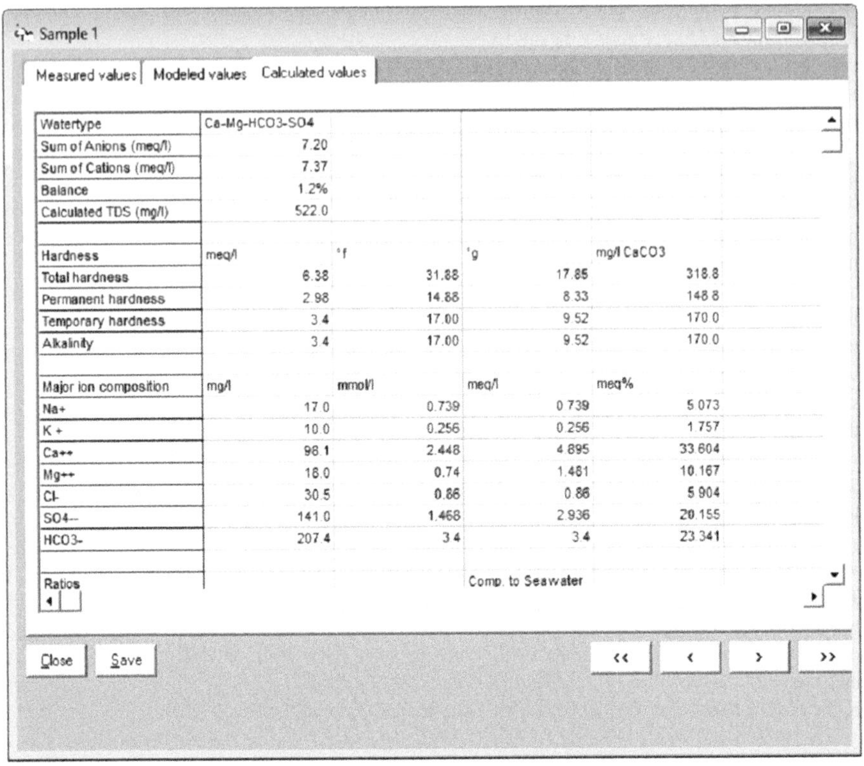

Sample 1

Measured values | Modeled values | Calculated values

Watertype	Ca-Mg-HCO3-SO4			
Sum of Anions (meq/l)	7.20			
Sum of Cations (meq/l)	7.37			
Balance	1.2%			
Calculated TDS (mg/l)	522.0			
Hardness	meq/l	°f	°g	mg/l CaCO3
Total hardness	6.38	31.88	17.85	318.8
Permanent hardness	2.98	14.88	8.33	148.8
Temporary hardness	3.4	17.00	9.52	170.0
Alkalinity	3.4	17.00	9.52	170.0
Major ion composition	mg/l	mmol/l	meq/l	meq%
Na+	17.0	0.739	0.739	5.073
K+	10.0	0.256	0.256	1.757
Ca++	98.1	2.448	4.895	33.604
Mg++	18.0	0.74	1.481	10.167
Cl-	30.5	0.86	0.86	5.904
SO4--	141.0	1.468	2.936	20.155
HCO3-	207.4	3.4	3.4	23.341
Ratios			Comp. to Seawater	

Close | Save | << | < | > | >>

Figure 9.5 *The calculated values for the sample shown in AquaChem.*

of virus that infects only the faecal bacteria *Escherichia coli*. This means that they will not have any detrimental impact on human or animal health if they end up in the water supply and makes them very attractive for use in groundwater tracing. The size of an individual phage virus is less than 0.2 μm, making them ideal for moving through very small pore spaces. The concentration of phages decays slowly with time, with a half-life usually measured in weeks.

Bacteria are rarely used as a tracer because of environmental objections, and the use of bacteriophages is preferred in most cases. Where a spring is already contaminated by a bacterium, however, it may be possible to use this as a tracer to identify the likely source of the contaminant. Once this has been found, the connection should then be confirmed using another tracer technique.

9.3.5 Isotopes

Both stable and unstable isotopes are used in groundwater tracer work and this is really a variation of the solute tracer method. The commonly used tracers are iodine-131, cobalt-60, bromine-82 and tritium (H^3). All radioactive tracers require the use of expensive analytical methods and careful sampling techniques. Unstable isotopes are rarely used because of general objections to introducing radioactive materials into the environment. Radioactive isotopes that occur naturally in groundwater, however, are used as tracers without any environmental objection. A detailed description of these methods can be found in the work of Davis *et al.*, (1985), Gat (1971) and Ward *et al.*, (1998).

9.3.6 Using groundwater tracers

Before using any form of tracer, permission must be obtained from the environmental regulator and local authorities responsible for public health and water quality. Complete a survey identifying all the possible locations where the tracer may emerge, including springs, gaining streams, wells and boreholes, so that the sampling points can be established. Also ensure that you alert local people to the possibility of a sudden change in the colour of their stream or water supply. Case History 5 shows how tracers may emerge in different locations and take a range of travel times to arrive.

You must choose the most suitable tracer for the test you are planning, and also the concentration needed when it is injected into the groundwater flow. Mechanical dispersion (see Section 3.3.3) is the process of spreading during flow and will have a significant impact on your experiment. Mixing and dilution may also rapidly reduce the tracer concentration which reduces rapidly below detection limits. Mechanical dispersion also affects the way that the tracer reaches the monitoring point. The concentration quickly builds up once the tracer has arrived, to reach a peak. This is followed by a very slow tail-off, with concentrations slowly decaying. This effect can be seen in the example shown in the Case History. The concentration of a phage will also naturally reduce as it travels through the ground, with the virus having no food source.

The time taken for the tracer to reach the monitoring point can be assessed using Darcy's law (see Section 3.5). The calculation will be approximate, so, to be on the safe side, monitor for twice as long before you pack it in if there has been no sign of the tracer. Tracers may not travel at the same speed as the groundwater flow and so an adequately long period must be allowed. Remember that if you fail to identify any tracer at the measuring point, it does not prove that there is no connection. The tracer may have been diluted below a detectable concentration, or particle tracers could have been filtered out, or dyes and solutes been adsorbed onto particles in the rock matrix.

As with all aspects of groundwater fieldwork, safety is paramount. A number of the substances used as tracers are potentially hazardous, and the appropriate precautions need to be taken to protect both yourself and others who may come into contact with them.

10

PRACTICAL APPLICATIONS

This chapter discusses the practical application of the techniques described in the earlier parts of the book to the most common types of investigation that hydrogeologists are involved with today. It is intended to get you started in the right direction, as it is not possible for a field guide to cover this work in full.

10.1 Borehole Prognoses

Before a new well is constructed, it is typical for a well driller to decide on the chance of success, and if the new well is to be outside the usual area where he operates or where he is uncertain about the geology, he may ask you for an opinion on the chance of obtaining a supply in that location. He may just email you with a grid reference and details of how much water is needed, ideally both as an instantaneous flow and on a daily basis. This is usually all the basic information you need to be able to give an opinion; in some cases, however, other information may be necessary such as data on other boreholes or a map showing the precise location of the proposed borehole.

For the purpose of this exercise I am assuming that you are in the UK, although the general principles of forming an opinion are the same wherever you are. The first thing to do is to find the site on a map. If you have a computer, you can get a map on sites such as www.streetmap.co.uk which allow you to look up the location by name or possibly by grid reference. This will tell you where the site is in relation to topographical features such as streams and rivers, hills and valleys, and will also tell you about the local area in terms of the density of housing and so on. It could be that satellite information using Zoom Earth, Google Maps and Instant Street View also help to get a picture of the site. This information will help you decide on the depth of the borehole and the amount of casing that is required. The latter decision also needs information on the geology, although the depth may be simply decided on by being some 10–30 m below the bottom of the nearby river valley, usually the major one.

Next you should consider the local geology by looking at the British Geological Survey (BGS) website (https://mapsapp2.bgs.ac.uk/geoindex/home.html) which shows the geology. In order to access local borehole records that are held by the BGS you need to log onto https://www.bgs.ac.uk/information-hub/borehole-records/. Unfortunately, although it is a legal requirement for all records of boreholes drilled for a water supply that are greater than 30 m deep to be sent to the BGS, not all well drillers do it, and the situation has got worse over recent years because the Environment Agency does not get involved at all where the abstraction rate is less than 20 $m^3 d^{-1}$. It appears that an increasing number of water well drillers ignore the requirement to provide these data and the BGS has no powers to enforce the legal requirement.

Using the geological information to decide on whether a successful borehole can be drilled is easy in theory but may be complicated in practice, as it depends on the detail available in the geology and on the aquifer properties. It is obvious that the quality of your advice depends on the detail of the geological information. The UK has been mapped at least once and most of it twice; the greatest difficulty lies with those small areas that were mapped only once, generally some 150 years ago, although the BGS website has the latest mapping on a 1:50,000 scale for all areas.

Field Hydrogeology, Fifth Edition. Rick Brassington.

If the transmissivity and storage are medium to high values, there should be no problem; the difficulties lie with low-permeability aquifers and so-called unproductive strata. These rocks do contain groundwater, albeit in small quantities. Use the Ordnance Survey or equivalent maps to identify springs from these strata and you may be able to get a small supply. However, do not be afraid to say that there is no realistic prospect of obtaining a supply when that is the case.

You can get information on local nature protection sites, source protection zones and many other features of the government map-based site Magic Maps (https://magic.defra.gov.uk/magicmap.aspx). When you log on you are faced with a screen that shows the whole of the British mainland from which you can change the scale and zoom into the area of interest. There is a list of the site contents on the lefthand side of the screen and by ticking the appropriate box you can get the local features under each heading.

10.2 Groundwater Supplies

Most groundwater abstractions are from boreholes, although many small supplies continue to be obtained from shallow wells and springs.

10.2.1 Constructing a new borehole source

A borehole can be thought of as comprising two parts: an upper section, which penetrates the material overlying the aquifer and the upper 3 m of the aquifer rock and is supported by well casing, and a lower section, which allows the water to enter. The upper section provides access to the lower section so that water can be pumped to supply. The Environment Agency recommends that the minimum depth of casing is some 13 m to take the borehole through the overlying materials and to penetrate the aquifer by at least 3 m – good advice, so make sure that you stick to it. The lower section may be an open hole with a rock wall, or have a slotted pipe to support it if the rock is not stable or when the borehole taps an unconsolidated aquifer. Driscoll (1986) and Misstear *et al.*, (2017) provide useful advice on borehole design; see also Section 9.1 of this book for an introduction to the drilling methods.

Develop a conceptual model for the locality and use it to decide on the borehole position, its total depth and the depth of casing. The diameter is determined by the pump size and construction needs and can be left to the driller, although do not forget that it needs to have a dip tube inserted, and possibly an extra one for a water level logger, so make sure that you remind the driller to make the borehole wide enough. The casing should be deep enough to provide support and keep out potentially polluted surface water or poor quality shallow groundwater. The need for slotted casing to support the lower section depends on the extent that the rock is fractured and can also be left largely to the driller. Screen and a gravel pack are only likely to be needed in sand or sand and gravel aquifers. Misstear *et al.*, (2017) provide a clear guide on this aspect of borehole design.

Make sure that the casing is grouted to prevent any leakage from the outside. Competent water well drillers will do this correctly, but beware of the cowboys who may ignore it altogether. The most reliable method for grouting casing is to drill the borehole to the depth of the base of the casing. The casing is then inserted, and a volume of cement-based grout sufficient to fill the space between the casing and the borehole wall and invade the geological materials, normally with an extra 20%, is inserted inside the casing. The casing is filled to the top with drilling mud and then sealed. The contents are pressurised by additional mud being pumped inside, and when the grout appears at the surface you know that the annulus is properly filled. Alternatively, make the cement bags into a ball, insert them in the borehole and push down with the rods. Methods using tremie pipes inserted into the annulus between the casing and the rock face are less reliable as it is difficult to get the grout all the way round the casing.

Only use the minimum quantity of water in cement grout or the grout will shrink and allow flow down the outside of the casing. Use three parts water to five parts cement by weight to ensure that the grout will remain liquid enough to pump and minimise shrinkage. This translates to 15 L of water to each 25-kg bag of cement. It is very important to add 1 kg of bentonite per 25 kg of cement, which will absorb surplus water and the expansion of the bentonite will avoid shrinking taking place.

Decide on the borehole depth using geological information on the aquifer thickness and depth. Make sure that the borehole is deep enough for the pump to always be well below the water level when it is running. An adequate depth is also needed to make sure that there is enough aquifer providing flow to the borehole. One of the most common faults with inadequate yields is due to the borehole being too shallow.

You may be asked to drill a borehole to prove the level of a fresh/saline groundwater interface. One way of doing this is to get the driller to take a water sample every time he puts on a new rod, assuming that he is drilling using an air flush. Measure the electrical conductivity (EC) of the samples and plot a vertical profile (Brassington & Taylor, 2012). Although the water sample will have a proportion derived from up the borehole, the majority will come from the location of the base of the drill rods. That will enable you to tell when the conductivity increases and hence the location of the interface, defined by the Environment Agency as the 1000 mg L^{-1} Cl isochlore. This is an oversimplification, as a real interface is a gradual increase in EC values over a vertical space that varies between 30 and 250 m.

Figure 10.1 shows an example of such a profile developed on a borehole at Warrington, Cheshire, England, with the profile built up from the samples shown in pink. It compares the profile with the

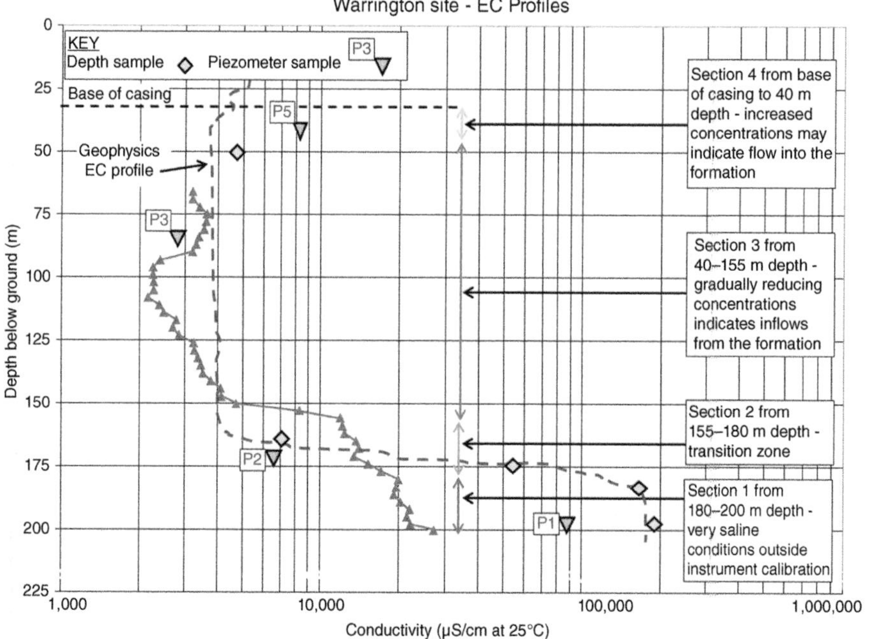

Figure 10.1 *A vertical profile of EC measurements on water samples taken at each rod change in the borehole. (Reproduced from Brassington F.C. and Taylor R. (2012) by permission of the Geological Society.)*

EC geophysical profile measured in the borehole (shown in blue) together with depth samples taken at the sample time as the borehole was logged. In addition, the diagram shows the conductivity of water samples taken from piezometers that all had 3-m response zones set at different depths. It can be seen that the EC profile generally matched the shape of the other data, albeit being slightly less than the more reliable final results. Results from a second site had similar results and, consequently, it can be concluded that this represents a quick and cheap method of identifying when a borehole is approaching a fresh/saline groundwater interface.

10.2.2 Investigating borehole performance

Borehole yield frequently deteriorates with time, caused by factors such as pump wear, interference from other sources and the well face becoming clogged. To investigate and quantify the performance of the well itself you should carry out a step test and produce a specific capacity curve, as explained in Section 6.7 and shown in Figure 6.6. If you have data from a step test carried out when the borehole was new, simply pump the borehole for several hours and check on the discharge rate. Do not rely on a water meter as its poor accuracy may be part of the problem you are investigating, so use a bucket and stopwatch. Measure the pumping-water level and then turn off the pump and allow the water levels to recover for a few hours, overnight or longer if possible. Calculate the drawdown and compare the yield and drawdown values with the specific capacity curve from the original step test. If the well performance has deteriorated, the drawdown value will have increased for any value of pumping (see Figure 6.6). If there was no earlier step test, carry out one now and use a Bierschenk analysis, as shown in Figure 6.11, to quantify the well losses. If the well losses account for most of the drawdown, it is a fair bet that the well face is clogged. Take a sample and utilise BART test kits, as described in Section 7.3.4, making sure that you use the three basic ones for iron-related bacteria, sulphate-reducing bacteria and slime-forming bacteria. If the borehole is 'contaminated' then it will need to be sterilised, and an experienced water well drilling contractor will be able to offer various methods of restoring the yield using oxalic acid or a proprietary brand of cleaning agent.

Where the water supply borehole contains dissolved iron and/or dissolved manganese these metals may be derived from minerals such as iron pyrite that are frequently associated with mudstones in sequences such as the Coal Measures in the UK. The oxygen that was dissolved in the recharge water has been lost to various reactions and so is able to dissolve these metals. The groundwater in the mudstone sequences can have relatively high concentrations in solution and this can cause problems with the water supply and may also clog the borehole walls. A good way to deal with this problem is to design the borehole so that the mudstone sequences are cased out and the only water entering the borehole is from the sandstone or limestone aquifers. However, this is not always possible and so the drilling contractor will install a treatment system to remove the dissolved iron or manganese from the water in supply. However, this does not treat the groundwater in the borehole where once the dissolved metals come into contact with oxygen that has been dissolved in the groundwater they precipitate on the borehole wall. This process is usually facilitated by bacterial action with a number of bacteria that live on the reaction of oxidation of iron and manganese. The effect of this precipitation is that the borehole well face becomes clogged and the borehole's specific yield is reduced. Specific yield is the stable pumping rate divided by the associated drawdown and is measured in $m^3d^{-1}m^{-1}$. To view the yield in this way gives a much more realistic view of a boreholes capability as it includes the drawdown.

The build-up of mineral(s) on the well face can be removed by treatment with the most popular chemical for this treatment being oxalic acid ($C_2H_2O_4$). This acid is available in liquid form with a trade name although it is also available in a solid form and comes in plastic sacks. The solid chemical looks like wet snow when the sack is opened as it is hygroscopic and absorbs water from the atmosphere and consequently the chmical doesn't blow about. However, it is a strong acid and should be treated with care when it is used. If you are planning to use it to treat a borehole you should first obtain a copy of the safety instructions and follow them.

Wearing a boiler suit, goggles, a mask, and gloves you can pour the chemical into the top of the borehole using a small bailer such as a coffee mug. Before you do this you should modify the top of the borehole so that the delivery pipe from the submersible pump is fed into the top of the borehole to allow recirculation to be carried out. You should calculate how much acid you need to put in the water to produce a strength of 5% when it is dissolved in the borehole water column. Once the acid is in the borehole switch on the pump and get it to circulate the water for a few hours. At least four hours is the minimum you should be aiming for although the longer the better as the agitation encourages the chemical to react and you could recirculate overnight if it is possible to arrange it. Then let the borehole rest for a few hours before you reconnect the pump so that the water can be pumped out to waste. If you are treating a small diameter supply borehole the volume will be fairly small and you will be able to catch it in a couple of Intermediate Bulk Containers (IBCs). Calculate the volume that you will be pumping out to remove three times the volume of the water stored in the borehole. Remember that an IBC holds 1 m^3 or 1,000 litres and so get hold of the number you will need. If you are treating a larger diameter borehole you will need a larger tank or more IBCs in which to store the water. Treat the water in the IBC by adding hydrated lime or another neutralising agent. Check the pH of the water before disposing of it safely.

You can check the effectiveness of the treatment by carrying out two short pumping tests, one before you treat the borehole and one afterwards. Pump for say, three hours and measure both the water level and the discharge rate in the usual way. Calculate the specific yield by dividing the pumping rate by the drawdown and you should get a higher value for the second test if your treatment was successful, i.e. the volume pumped for a given drawdown should be bigger.

10.2.3 Investigating abstraction impacts

The sustainability of a groundwater abstraction depends on the available recharge to the aquifer and the existing 'commitments' from these resources to existing users and to maintain surface water flows and wetland habitats. Use the water balance approach to identify the total resources and then calculate the existing total abstraction by identifying the local sources. The requirements to sustain wetlands will depend on the dry weather flow rates of the potentially affected streams and the potential for lowering groundwater levels in the vicinity of wetland sites.

To examine the potential impacts, you will need your conceptual model that defines the flow system, data from pumping tests to quantify the extent of drawdown, and field surveys to locate and identify local wetlands, ponds, and water sources that may be affected. A guide to select radius for the search is given in Table 6.4. Use the conceptual model to ask yourself the simple question 'Where would the water have gone if the new borehole did not exist?' Remember that water abstracted from a borehole has been diverted from its natural flow path and your investigation should be directed at identifying the parts of the flow system that will be depleted, quantifying the volumes of water involved and the significance of the impacts that will result.

Groundwater-level data taken during a pumping test on the new borehole, other boreholes nearby or a piezometer close to a wetland will help identify and quantify any impacts. Plot the drawdown data on a semi-log graph as for a Cooper and Jacob analysis (see Section 6.10.5), with the time scale extended so that the drawdown can be predicted after 1 or 2 years of constant pumping, as shown in Figure 10.2. The graph shows that the rate of decline in the water level increased after about 1600 minutes (1.1 days). To assess the long-term drawdown, the line that represents the rate of decline at the end of the test was extended to 1,000,000 minutes (almost 2 years) and predicts a drawdown of just less than 18 m. This prediction is conservative, as it both assumes continuous abstraction at the rate of the pumping test and ignores recharge, so the predicted drawdown is very much a worst case.

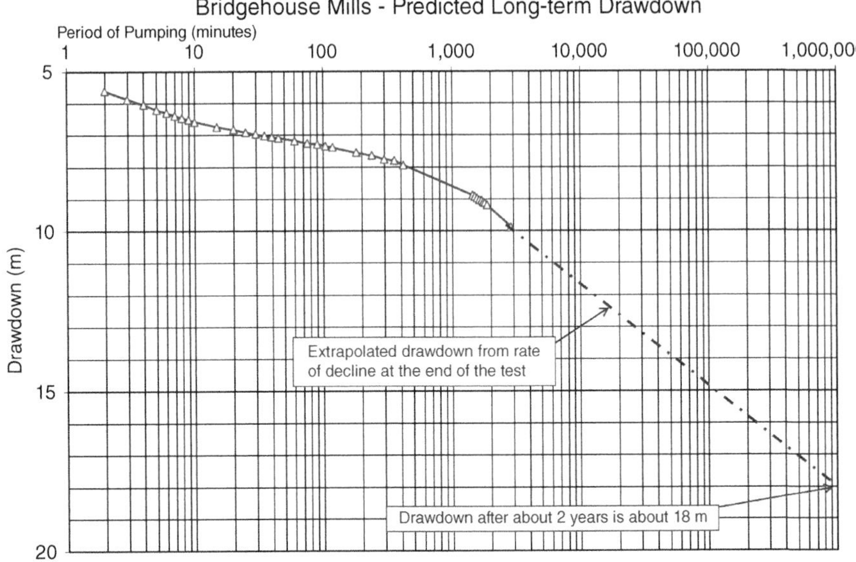

Figure 10.2 *During a pumping test on a borehole in a sandstone aquifer, water level measurements were made on the borehole. Drawdown data were plotted on a semi-log graph as shown, with the line extended to 1,000,000 minutes (almost 2 years) to predict a long-term drawdown of 18 m.*

The distance over which a pumping borehole may lower water levels can be assessed using a distance–drawdown semi-log graph (see Section 3.5), with an example seen in Figure 3.12. This analysis suggests that stream and spring flows inside this zone could be depleted in the long term. However, it should be remembered that direct measurements over the relatively short period of a typical pumping test depletion only very rarely detect the early signs of these long-term effects. Such impacts are normally anticipated by examination of the groundwater-level data as discussed here, or identified by long-term monitoring over a period of years during the early life of the source.

10.2.4 Spring supplies

Spring flows are liable to be very variable and therefore their reliability as a water source can only be assessed by flow measurement, using the techniques discussed in Section 5.3. Measurements in late autumn or early winter, when groundwater levels are expected to be at their lowest, will help define the minimum flows. Many springs react quickly to recharge from storm events, so take care that you use the dry-weather flow values in your reliability assessment.

Most springs drain shallow aquifers and are more vulnerable than borehole sources to contamination from a number of microbes, including *Cryptosporidium*. This is a protozoan that infects humans and animals such as cattle and sheep. Over recent years it has become common in private water supplies and it is strongly recommended that all such supplies are treated with ultra-violet light that destroys bacteria, viruses, and protozoa. A spring source should be protected from pollution by constructing a catch pit (see Figure 10.3). Water is taken into supply under gravity flow or may be pumped with the intake pipe fitted with a filter screen. An overflow controls any excess water, and a drain and inspection cover are provided for ease of maintenance. The inspection cover

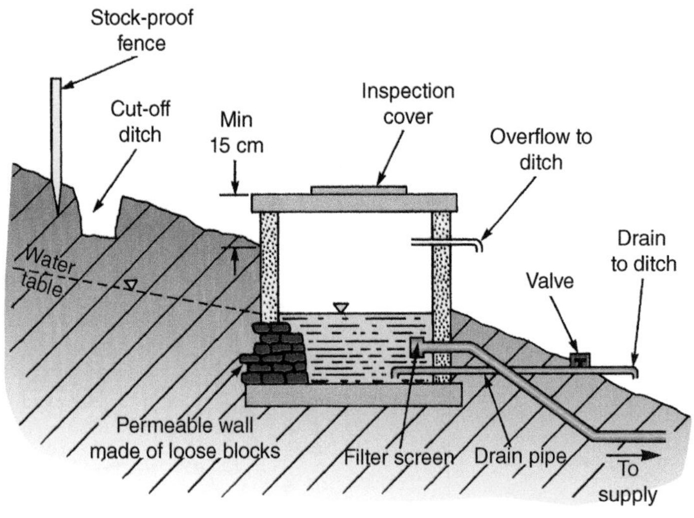

Figure 10.3 Protect a spring source with a catch pit that is large enough to capture the entire flow and incorporate the features shown here.

should be lockable and waterproof, and the overflow pipe should be made vermin-proof to prevent animals or insects from polluting the supply. A cut-off ditch runs above the spring to divert surface water from the chamber, and animals are kept away by a stock-proof fence.

Pollution problems should be anticipated from activities in the catchment, such as sewage sludge spreading, the use of agricultural fertilisers and pesticides, and the disposal of agricultural waste such as silage liquor or spent sheep dip. In fissured aquifers, especially karstic limestones, or where there is a steep topography, groundwater flow is rapid and pollution of spring supplies can happen quickly after noxious materials are spilt or deliberately spread onto land. It is prudent to carry out a thorough investigation, including tracer studies, to establish the vulnerability of the supply to pollution hazards.

10.3 Wells in Shallow Aquifers

Where an aquifer is very shallow it may not be possible to construct a successful 'normal' borehole, because there needs to be enough vertical space for a submersible pump and the inlet screen (perforated casing), which is preferably located below the pump. Since the pumped water level should be above the submersible pump to keep it cool, drawdowns are severely restricted in shallow unconfined aquifers and the situation is made more difficult by seepage faces. As a result, a lot of opportunities for a water supply are missed despite the aquifer containing a great deal of water.

The simplest way to abstract is from a large-diameter (3 or 4 m) well dug to the base of the aquifer or at least deep enough to be able to get a supply. When these wells are pumped a great deal of the supply is obtained from well storage and groundwater will continue to flow into them after pumping has ceased; this can be as much as 70% of the total yield of the system (Rushton and Holt, 1981). However, these are not always popular as they require a lot of land and maintenance. Alternative methods are vertical wellpoints and horizontal wells, as described by Rushton and Brassington (2013a, 2013b, 2016) and Brassington and Preene (2003).

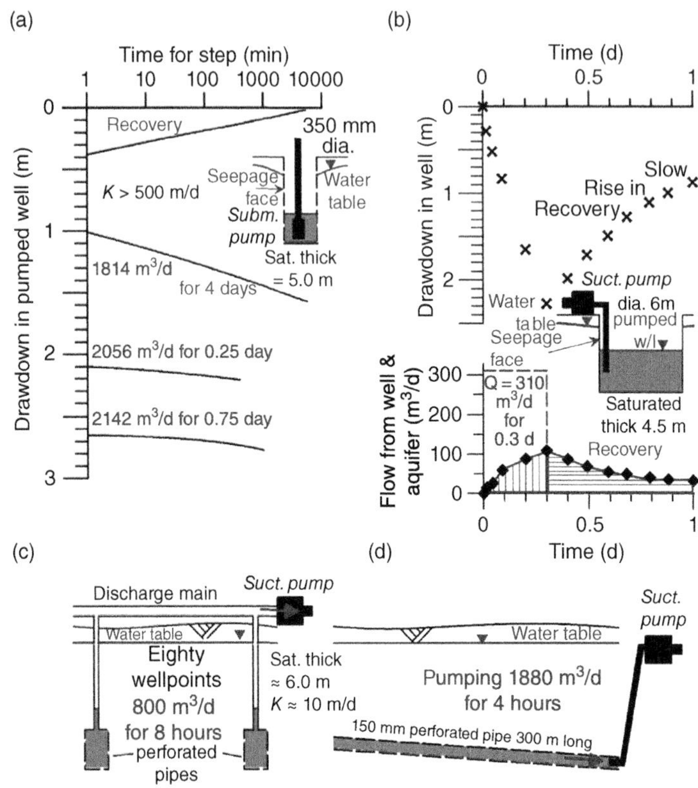

Figure 10.4 *Field examples of exploitation of an unconfined aquifer of limited saturated thickness: (a) vertical well with submersible pump in gravel aquifer, (b) large-diameter well in weathered zone, (c) wellpoints in sand dune aquifer, (d) horizontal well in sand dune aquifer. (Reproduced from Rushton and Brassington (2016) with permission of Copyright © 2015, Springer-Verlag Berlin Heidelberg.)*

Wellpoints are small-diameter (often about 50 mm) wells that are inserted in lines and separated by about 2 m. They are usually used for dewatering sites where civil engineering works are being carried out, although they may be used as a water source where the aquifer is too shallow to allow a borehole to be constructed. The problem is that they are designed principally for dewatering and so may not last as long as a horizontal well.

Horizontal wells can be constructed using a deep trenching machine with a chain of buckets that cuts a temporary trench and inserts a land drainage pipe at depths up to 6 m. The land drainage pipe is usually made of corrugated PVC with a diameter of 160 mm and a slot size of 2 mm and is factory-wrapped in Terram 1000 geotextile with a pore size of 150 μm. Typically such installations are about 300 m long and have a filter sand material inserted partly round them as they are constructed through a special funnel on the trenching machine. Yields of up to 70 $m^3 h^{-1}$ or more have been sustained from such installations that are pumped for just a few hours.

Perhaps the most difficult aspect is finding a contractor with a trenching machine capable of excavating to 6 m depth, as most such machines are used to install field drains they are restricted to some 2 m and it is now known there is only one such machine owned by a UK contractor. Alternative methods of construction involve deep trenching using traditional methods that probably will require temporary dewatering using a wellpoint system (Mailvaganam *et al.*, 1993). Figure 10.4 illustrates the main features of abstraction systems from shallow aquifers.

10.4 Contaminated Land Investigations

Contaminant hydrogeology has been the aspect of the science that has seen the greatest growth since the early 1980s, in terms of both the number of hydrogeologists working in this field and the number of projects that need their attention. There are many potential sources for the pollutants found in groundwater, such as leaking sewers and pipes, chemical spills, uncontrolled waste disposal and the agricultural use of fertilisers and pesticides. Studies of groundwater pollution are often associated with contaminated land investigations, as the groundwater flow system usually provides the most effective flow path for pollutant migration. Several textbooks, such as those by Bedient *et al.*, (1999) and Fetter (2018), provide details of this type of investigation.

In almost all groundwater pollution investigations the source–pathway–receptor model or concept is used. The source is the source of the polluting material, the pathway is normally the groundwater flow system, and the receptor is likely to include water sources or surface water bodies. In some cases, the groundwater body itself is regarded as the receptor. For example, any substance included on List I of the EC Groundwater Directive (Council Directive 80/68/EEC on the Protection of Groundwater Against Pollution Caused by Certain Dangerous Substances) cannot be discharged into groundwater. List I substances include most pesticides and herbicides, most solvents, mineral oils and hydrocarbons, cadmium, and mercury.

10.4.1 Groundwater contamination

The pollutants in groundwater can be divided into three types according to their behaviour, and comprise solutes, light non-aqueous phase liquid (pronounced 'el-napple', LNAPLs) and dense non-aqueous phase liquid (pronounced 'dee-napple', DNAPLs). Soluble substances will travel with the groundwater flow as a solute plume that will vary in concentration across it. Non-soluble and poorly soluble liquids are termed non-aqueous phase liquids (NAPLs); these are either less dense than water (LNAPL) or more dense (DNAPL). LNAPLs float and travel with groundwater flow. They are encountered in boreholes as a liquid floating on the water surface, as a vapour and in solution in very low concentrations. DNAPLs, on the other hand, move through the aquifer mainly under the influence of gravity and accumulate at the bottom of the aquifer, or form local concentrations on low-permeability layers and do not travel with the groundwater flow. This makes them very difficult to detect. The value of the aquifer's permeability in respect of NAPLs will be different to the hydraulic conductivity, usually resulting in slower flow rates for LNAPLs and higher flow rates for DNAPLs due to the different properties of the liquids involved. The interpretation of contaminant movement in groundwater systems is complex, but provided you do not panic and apply hydrogeological common sense you are likely to reach the right conclusions.

Groundwater contamination is very difficult to remove and it is generally accepted that it is not practical to restore the groundwater to its original condition. Instead, clean-up programmes are designed to remove enough of the contaminant to render the groundwater acceptable for abstraction, provided that the water is treated prior to use. The most usual clean-up method is 'pump and treat', where the contaminated groundwater is extracted through wells and treated to remove the main contaminants before being recharged back to the aquifer. It often takes many years to reduce the contaminant concentrations to acceptable levels, because as the treatment reduces concentrations in the groundwater, greater volumes of water have to be pumped through the treatment equipment to remove the pollutants. Surface tension forces make it difficult, if not impossible, for all the contaminants to be removed in this way.

Bio-treatment methods use naturally occurring bacteria to reduce the concentration of pollutants and have been in use since the 1990s. Bacteria present in the contaminated groundwater or soil are identified to see whether any use the pollutant as a food source. They are then grown in large quantities and sprayed onto the ground or injected into the aquifer. The theory is that they will flourish until all the contaminant is used up. Once the food source has gone, the numbers of bacteria will quickly decrease to natural background levels. Other forms of treatment involve using strong oxidants such as hydrogen peroxide that will break down the more complex compounds into their basic molecules and hydrogen peroxide breaks down to oxygen and water.

10.4.2 Investigating contaminated land

Most industrial activities contaminate the site where they are located. The scale ranges from huge factories to small back-street operations. Industrial activity has been undertaken in most countries since the Industrial Revolution commencing in the mid 18th century and this means that much of old urban areas is largely built on contaminated land. Such areas are a potential source of polluting groundwater, either by material being leached out of the ground or by rising water tables flooding it and providing a pathway for the pollutants to escape.

The investigation of contaminated land is a wide and complex topic that is outside the scope of this field manual. The environmental regulators such as the Environment Agency in England, Natural Resources Wales, the Scottish Environment Protection Agency, the European Environment Agency, and the United States Environmental Protection Agency have guidance notes to assist in the understanding of local regulations controlling the investigation of contaminated land. A good starting place to obtain copies of such publications is to look at the websites for these and similar organisations.

Work on contaminated sites is potentially hazardous and you should take careful precautions in all fieldwork you undertake. This includes wearing appropriate protective clothing when carrying out field work and when handling contaminated samples.

10.5 Landfills and Leachate

Waste disposal is tightly controlled in most countries. EU member states, for example, are required to enforce national regulations that meet the requirements of the European Union Landfill Directive (Council Directive 1999/31/EC on the Landfill of Waste). These regulations stipulate that all sites must have a hydrogeological assessment undertaken and what it should include. The environmental regulators in all countries provide guidance on what is required, so obtain a copy of the appropriate guidance before undertaking this work. The volume of materials that is now going into landfills has dropped very significantly due to recycling and, consequently, hydrogeological work associated with landfills may now concentrate more on ceasing landfilling rather that continuing to deposit waste materials.

Landfills are regarded as engineering structures designed to prevent all leakage of any pollutants and to minimise the amount of rainfall that soaks into them. Leakage occurs, however, and monitoring is needed to identify when it happens so that appropriate remedial action may be taken.

10.5.1 Investigating potential cemetery sites

In the UK, planning permission for new cemetery sites requires an investigation to show that problems of water contamination will not occur. The assessment requires understanding the geology, identifying potential aquifers and monitoring groundwater levels. It is necessary to define the water table across the site in terms of its depth below ground level and the relative levels that will allow the flow direction to be determined. Permission will only be given if the water table is always more than 1 m below the bottom of the graves, which is taken to be the same as 3 m below ground level.

The Environment Agency (2004) has published a guidance note that includes the potential contaminant released from a single 70-kg burial over a 10-year period (Table 10.1), which shows that after a decade all the parameters will become zero except for the total organic carbon, which is just above the detection limit at that time. With continuing burials each year, a steady state will develop, with concentrations of each parameter equalling the annual totals over 10 years per burial. The total mass for each parameter is shown at the bottom of Table 10.1 in kilograms. The concentration in the groundwater (in milligrams per litre) will be the total loading (in kilograms) divided by the total annual groundwater flux (in cubic metres). The source–pathway–receptor model is then used to calculate the anticipated loading on all significant receptors such as a water source or a spring at the head of a watercourse.

At the site described in Case History 2, six piezometers were installed and monitored for a year at the sites shown in Box Figure 2.1. The hydrographs were then compared with an existing long-term record for a local observation borehole in a similar aquifer to calculate the highest water table elevation expected at the site (see Figure 10.5). Monthly readings in the six piezometers at the site shown in Box Figure 2.1 were used to construct the hydrographs in the upper graph (Figure 10.5). The amplitude of the winter 2004 peak above the level at the end of the recession in late December 2003 varied from 0.46 m to 0.75 m. The amplitude for this peak in the observation borehole record (Figure 10.5) was 0.503 m above the measurements at the end of the recession. The highest recorded groundwater levels in England were in December 2000, when the peak in this record was 0.89 m above the same datum. The December 2000 peak for the proposed cemetery site is calculated by applying the ratio between these values (0.89/0.503 = 1.77) to the first data set. The highest level (Piezometer No. 1) will be $0.75 \times 1.77 = 1.33$ m, which will be an elevation of 114.9 mOD. In Piezometer No. 5 this calculation is $0.6 \times 1.77 = 1.06$ m with an elevation of 114.7 mOD. These levels represent a depth of 3.53 m at Piezometer No. 1 and 3.79 m at Piezometer No. 5, and, consequently, under the highest groundwater conditions the water table will be more than 3 m below ground.

The total annual loading has been calculated assuming an average of 20 burials per year using the 10-year totals from Table 10.1; see lines 1 and 2 in Table 10.2.

The Meteorological Office long-term average annual effective rainfall for grass in this area is 214 mm. All the effective rainfall is expected to infiltrate through the grass surface that covers the site, and therefore the total recharge over the 2-ha site will be $20\,000 \times 0.214 = 4280\ m^3\ year^{-1}$. If it is assumed that the leachate from the graves will percolate to the water table and then be carried in

Table 10.1 *Potential contaminant release in kilograms from a single 70-kg burial.*

Year	TOC	NH_4	Ca	Mg	Na	K	P	SO_4	Cl	Fe
1	6.00	0.87	0.56	0.010	0.050	0.070	0.250	0.210	0.048	0.020
2	3.00	0.44	0.28	0.005	0.025	0.035	0.125	0.110	0.024	0.010
3	1.50	0.22	0.14	0.003	0.013	0.018	0.063	0.054	0.012	0.005
4	0.075	0.11	0.07	0.001	0.006	0.009	0.032	0.027	0.006	0.003
5	0.37	0.05	0.03	<0.001	0.003	0.004	0.016	0.012	0.003	0.001
6	0.19	0.03	0.02	<0.001	0.002	0.002	0.008	0.006	0.002	<0.001
7	0.10	0.01	0.01	<0.001	0.001	0.001	0.004	0.003	<0.001	<0.001
8	0.05	<0.01	<0.01	<0.001	<0.001	<0.001	0.002	0.001	<0.001	<0.001
9	0.02	<0.01	<0.01	<0.001	<0.001	<0.001	0.001	<0.001	<0.001	<0.001
10	0.01	<0.01	<0.01	<0.001	<0.001	<0.001	<0.001	<0.001	<0.001	<0.001
Total	11.315	1.73	1.11	0.019	0.1	0.139	0.501	0.423	0.095	0.039

Based on the Environment Agency (2004) booklet *Assessing the Groundwater Pollution Potential of Cemetery Developments.*

(a)

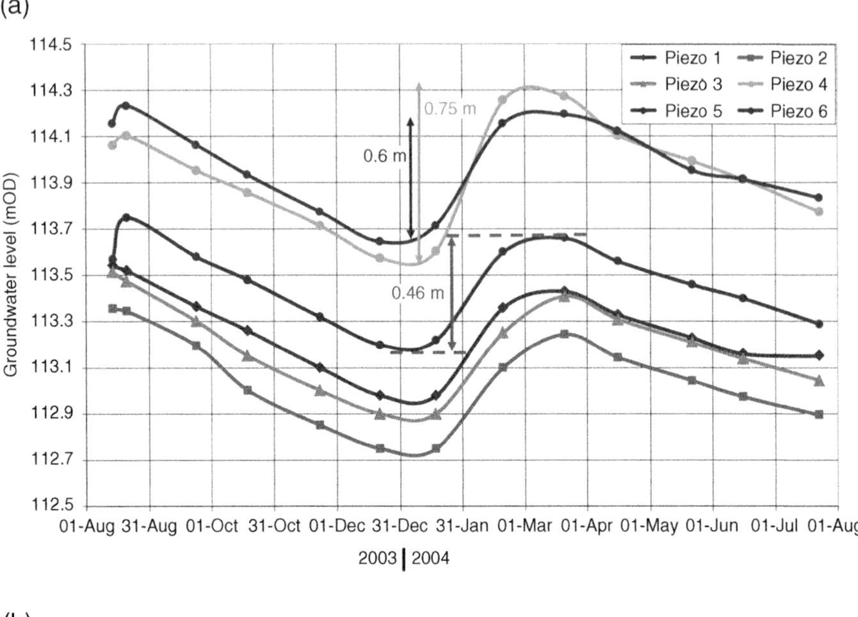

(b)

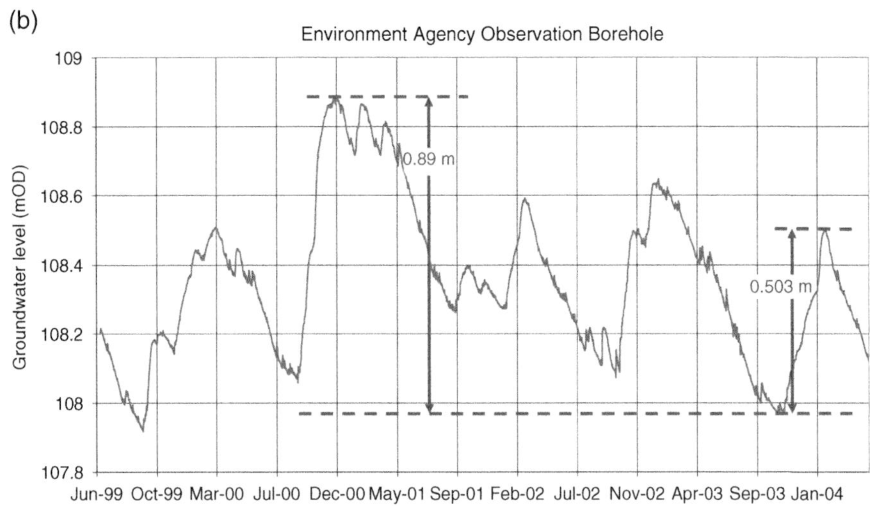

Figure 10.5 *A data set can be extended by comparison with a longer data set. This method is used here to determine the expected highest groundwater levels at the proposed cemetery site.*

the groundwater flow, the whole of the recharge will be available as dilution water at the western boundary of the site. This produces the concentrations shown in line 3 of Table 10.2.

The water table contours show that the groundwater flows beneath the adjacent fields towards the sandstone aquifer. Further dilution will occur from recharge through these fields. The distance is similar to the length of the site and the flow lines are expected to be parallel. Consequently, the area

Table 10.2 *Calculated contaminant loading.*

	TOC	NH_4	Ca	Mg	Na	K	P	SO_4	Cl	Fe
Steady-state loading (kg) per 70-kg burial	11.315	1.73	1.11	0.019	0.1	0.139	0.501	0.423	0.095	0.039
Total mass (kg) for 20 burials per year	226.30	34.60	22.20	0.38	2.00	2.78	10.02	8.46	1.90	0.78
Concentration (mg L^{-1}) at site boundary	52.874	8.084	5.187	0.089	0.467	0.650	2.341	1.977	0.444	0.182
Concentration (mg L^{-1}) at sandstone boundary	26.437	4.042	2.593	0.044	0.234	0.325	1.171	0.988	0.222	0.091
Concentration (mg L^{-1}) at water source	0.137	0.021	0.013	0.000	0.001	0.002	0.006	0.005	0.001	0.000

providing the additional recharge is about the same as the site and will halve the concentrations to the values shown in line 4 of Table 10.2 before it reaches the sandstone. The groundwater from the cemetery would then join the groundwater flowing towards the water supply source. By the time it reaches the abstraction point it will be further diluted by a factor equal to the ratio of the overall volume leaving the cemetery to the total abstraction from the wells. This factor is 193 and produces the concentrations shown in the bottom line of the table. If these values are compared with Table 7.11 it can be seen that the cemetery does not represent any threat to the water supply source. These calculations can be regarded as conservative because they ignore any reduction in the concentrations by processes other than dilution.

10.6 Geothermal Energy

There is a lot of public interest in ground source heat at the moment and some hydrogeologists are involved in its production. Banks (2012) explains in detail the theory and practice of utilising this source of energy.

Ground source heat is the mundane form of heat that is stored everywhere in the ground at normal temperatures. It is usually taken as being low-enthalpy heat in relatively shallow ground defined as depths of about up to 200 m and temperatures of less than 30°C although temperatures of up to 90° C may still be considered to be low enthalpy.

Ground source heat can be extracted and used in two general systems: open-loop and closed-loop. In an open-loop system water is pumped from one borehole, put through heat exchangers and then discharged back down a second borehole without being exposed to the air. Heat is extracted from the pumped water or, in a cooling mode, is dumped into it. In a cooling mode it is not necessary to use a heat pump, as the cooling water can be pumped through a network of heat exchange elements in a building to provide passive cooling. However, it is usual to use a heat pump to provide heating or active cooling. It is possible to abstract water from any source such as a river or stream or even horizontal wells. In the closed-loop system the borehole has a long length of small-diameter piping installed in it and is usually backfilled using a thermally conductive grout. A refrigerant is pumped through the small-diameter pipe and it transfers heat either to or from the ground depending on the relative temperatures.

It is far more difficult to get water into the ground, so open-loop systems frequently only work where there is sufficient head to drive the water into the ground, which means a relatively deep water table. It is also important to keep the groundwater away from the atmosphere so that oxygen does not dissolve in the groundwater. This will cause any dissolved iron in the groundwater to precipitate as ochre and block pore spaces, reducing the capacity of the discharge borehole to receive the water. Alternatively, the water can be discharged to a local watercourse, although that may mean that the abstraction has a greater local environmental impact and the scheme may not be permitted.

10.7 Groundwater Lowering by Excavation

Water supply abstraction is not the only cause of lowered groundwater levels. Quarrying, mining, and civil engineering works can all cause significant long-term or even permanent lowering. When a large excavation is made into an aquifer, it increases the void space (i.e. effective porosity or specific yield) to 100% in an open void. In clean gravel aquifers, this represents an increase of three- or four-fold, but in some dense aquifers such as limestone the increase can be as much as a hundred-fold. Groundwater will flow to fill this extra space, thereby lowering the water table in the vicinity (see Figure 10.6).

A long-established quarry that was excavated into a Namurian Sandstone with all the rock removed except for about 1 m at the base across the width of the quarry is shown as a plan in Figure 10.7a. A series of monitoring boreholes showed hydrographs that appeared to be different until they were examined in more detail, as can be seen in Figures 10.7b and 10.7c which show hydrographs from boreholes above and below the site. Those above the site appear to have no movement, although when looked at in detail they have the same fluctuations as those below the quarry, but with a much reduced amplitude. With this quarry there were no water supplies affected; however, you may find yourself involved in investigating a loss of yield or the causes of a dried-up wetland that result from one of these activities. Use the concepts outlined here to help plan your investigation.

The amount of lowering depends upon the depth and total volume of the excavation below the water table and the aquifer's hydraulic conductivity and specific yield. Groundwater will flow into the excavation until the former water levels are achieved and the original groundwater flow regime is re-established. However, when an excavation is dewatered by pumping, both the period of

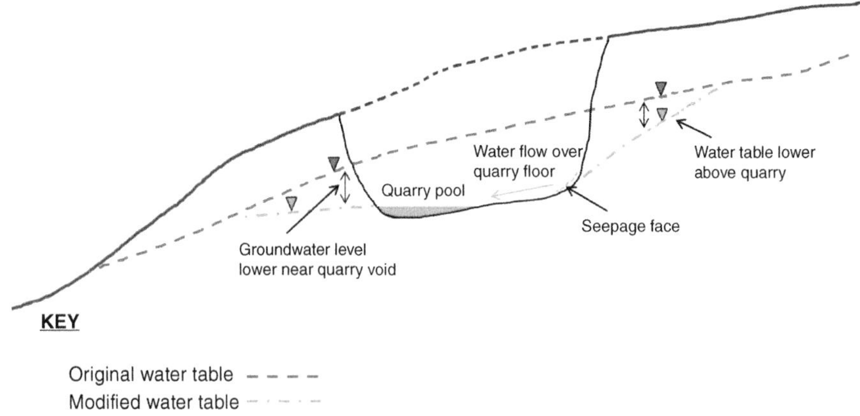

Figure 10.6 *Before and after conditions show how a large excavation into an aquifer can lower the water table by increasing the available aquifer storage. The impact may be greater if the quarry is dewatered.*

disruption and the extent of the area affected will be extended. Permanent lowering of groundwater levels can occur if the quarrying has the effect of moving a spring line. Use information from the original investigation boreholes to predict the final recovery level when a quarry is allowed to flood. Where the specific yield is lower and the quarry is large it may take several years for the recovery to reach the natural water table. If the quarry is on a hillside and near the top of the hill it may mean that there is not sufficient recharge to allow the water level in the quarry to recover, with water being lost to evaporation and flow through into the aquifer downstream.

Almost all deep mines are drained by pumping or gravity drainage through drainage adits (soughs), because either the workings are in water-bearing rocks or the shafts penetrate aquifers

(a)

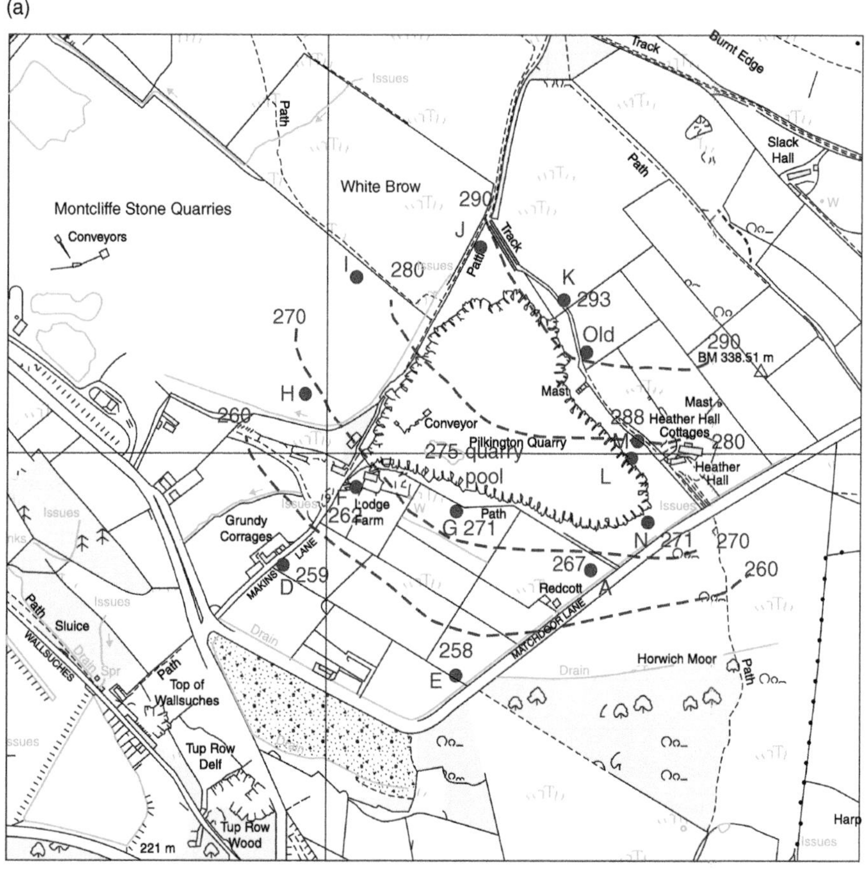

Figure 10.7 Monitoring boreholes around a sandstone quarry are shown in (a), together with groundwater contours in metres above Ordnance Datum (approximately sea level). In (b) the hydrographs are seen, with those below the site having a greater fluctuation than those that lie above it where the fluctuations appear to be almost non-existent. The actual fluctuations can be seen in (c) when the hydrographs are plotted on a larger scale. The difference between the hydrographs is explained by the presence of the large quarry void that pulls down the water levels above the site and does not affect those below it so much.

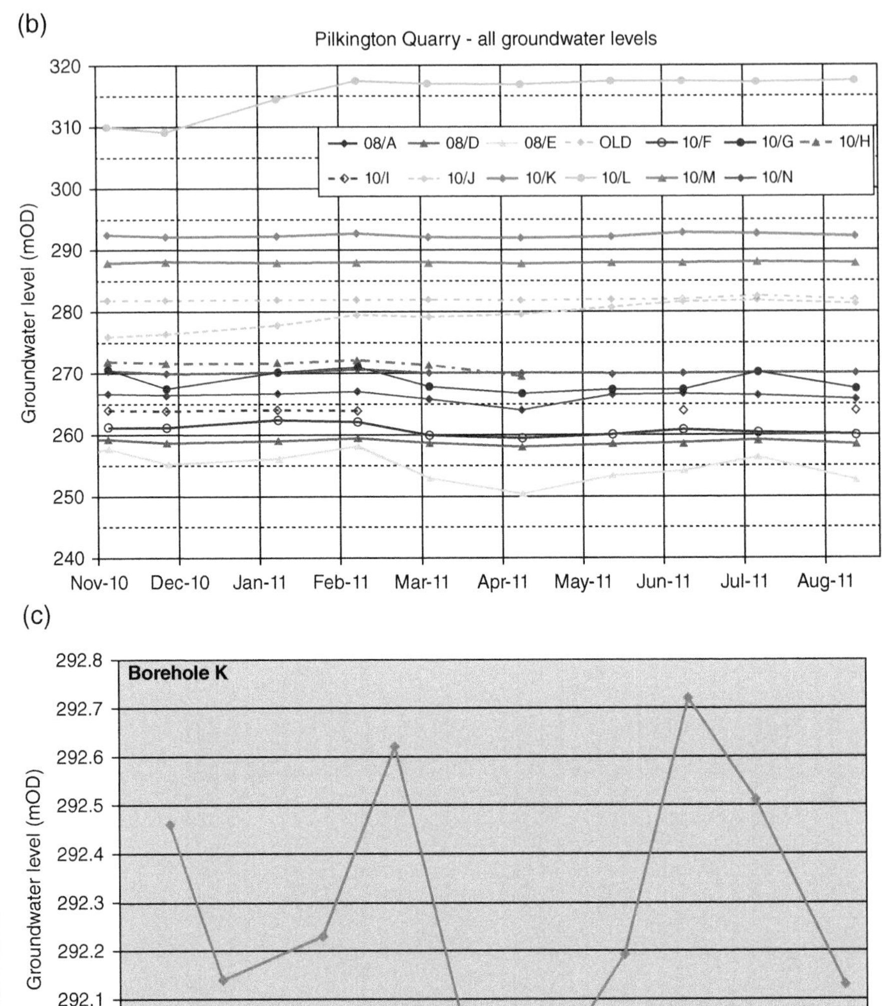

Figure 10.7 (*Continued*)

above them. Consequently, large volumes of water drain continuously from deep mines to cause dewatering over extensive areas. They also change the groundwater flow pattern by providing new flow paths via the workings and drainage adits. Subsidence resulting from mine workings may cause fracturing which can increase permeability on a local scale. Groundwater levels recover (often called rebound) when mining finishes and may take decades to complete, eventually causing localised flooding or the emergence of polluted groundwater discharges.

Many construction projects involve short-term excavations, some of which are below the water table. These have a similar impact to quarrying, although usually with a much shorter duration, which limits the impact on groundwater supplies. Permanent drainage systems are incorporated into some works, either to prevent groundwater pressures building up, which may alter the stability of slopes, the bearing properties of rocks and soils, or to prevent the flooding of cuttings and tunnels. Such drains usually run along both sides of cuttings and discharge into local streams where the topography allows. The drains allow groundwater to flow along them, thereby causing permanent depletion of groundwater levels. In the example shown in Figure 10.8, the cutting has reduced the area of the spring catchments by more than 50%, thereby halving the spring flows.

New tunnels or even major pipelines can have the same result, with groundwater flowing outside the tunnel lining or along the pipe bedding material. A survey to identify those sources at risk should be undertaken before work on the projects starts. A monitoring programme started before the work begins will mean that any impacts can be identified. Groundwater flow along backfilled pipeline trenches can be reduced by constructing anti-seepage collars around the pipe. These are a series of short cuts at right angles across the main trench and at least five pipe diameters in length and to a similar depth. They are backfilled with mass concrete to near ground level. A series of half a dozen should be installed at intervals equal to ten pipe diameters along critical lengths of pipe.

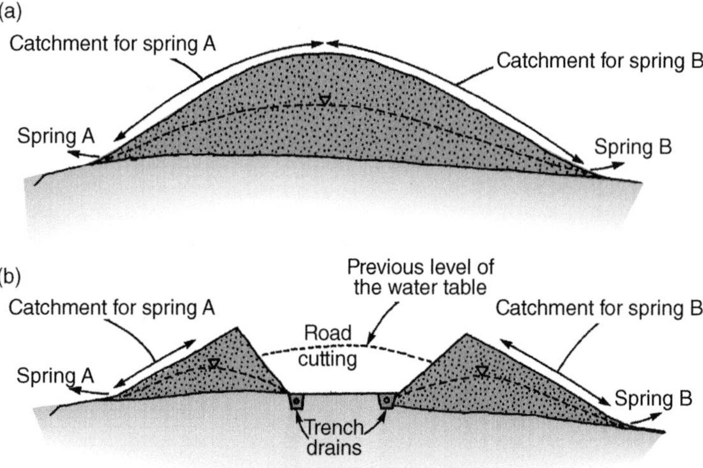

Figure 10.8 *A road cutting has been excavated through a hill formed from an aquifer with springs that discharge at A and B (a). Once the cutting was excavated below the water table, it was kept dry by groundwater being diverted along trench drains installed on both sides (b). The flow of both springs is approximately halved with their catchment areas reduced by about 50%. (Note: ditches are usually installed at the top of cuttings to help maintain slope stability, but these have been omitted from the diagram.)*

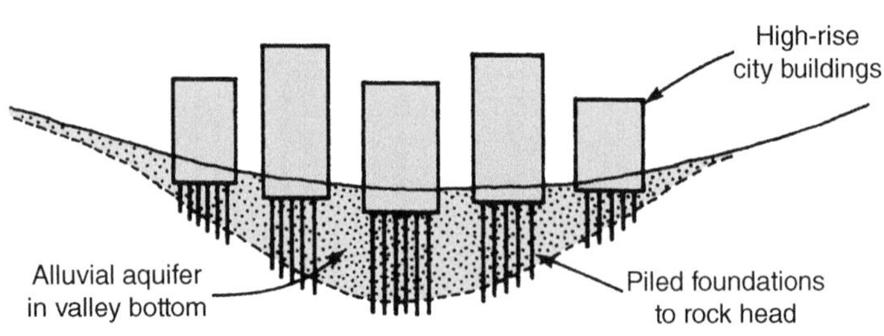

Figure 10.9 *The piled foundations of high-rise buildings may be sufficiently dense to reduce the effective cross-sectional area of a shallow aquifer, thereby reducing transmissivity. This will cause groundwater levels to rise on the 'upstream' side of the buildings, and alter the flow pattern round it which might affect the yield of nearby wells.*

Piled foundations, especially in urban areas, can radically reduce the cross-sectional area of shallow aquifers and may result in groundwater level changes (see Figure 10.9).

In shallow aquifers, the deepening of ditches or streams and the installation of a field drainage system will usually lower the water table in the immediate area. Figure 10.10 shows how this can happen. Tile drains (Figure 10.10a) lower the water table, which may cause a shallow well to become dry. A similar effect is seen in Figure 10.10b, where a ditch has been deepened. In Figure 10.10c, a spring line is thrown out over clay that overlies an aquifer. After the clayey field has been improved by the installation of tile drains (Figure 10.10d), the groundwater flow is diverted to the drain, causing the spring to dry up. In most cases the cure is to deepen the affected wells or replace spring supplies with a well or borehole.

10.8 Rising Water Tables

Incidents of rising water tables or groundwater flooding are now quite common in many countries, flooding basements in buildings, and other deep structures such as railway tunnels, or even areas of low-lying land (Wilkinson and Brassington, 1991). The usual cause is reduced groundwater pumping from either groundwater abstractions for water supply or the cessation of deep mining. Both need investigations by studying the relevant records. Leaking water mains and sewers occur in older urban areas and in some instances have been sufficient to cause the water table to rise. Other causes have included the excessive irrigation of crops and even parks and gardens, which can be identified by examining irrigation records and comparing the rate of application with expected transpiration rates.

Often, rising groundwater levels are a popular scapegoat for flooded basements that actually result from leaking water mains or sewers. Of all the many such incidents that I have investigated, all have been caused by leaking pipes of one type or another. Look for evidence of groundwater levels from nearby boreholes and test the chemistry of the floodwater, including the presence of free chlorine and THMs (see Section 9.3.2). Check for faecal bacteria, although diluted sewage may not contain any if the site is more than about 50 m from the leak. Sewage may cause increases in the nitrate, nitrite and ammonium content and can also result in increases in chloride. One substance that can prove a sewer connection is caffeine, which can be detected by a laboratory. Coffee and tea both contain caffeine, as well as a large number of soft drinks, and this substance makes its way into the sewer via the human system after it has been consumed. It does not naturally occur in groundwater and so, if you find it, it will establish a link. It is moderately soluble in water at room temperature (2 g/100 mL) and very soluble in boiling water (66 g/100 mL).

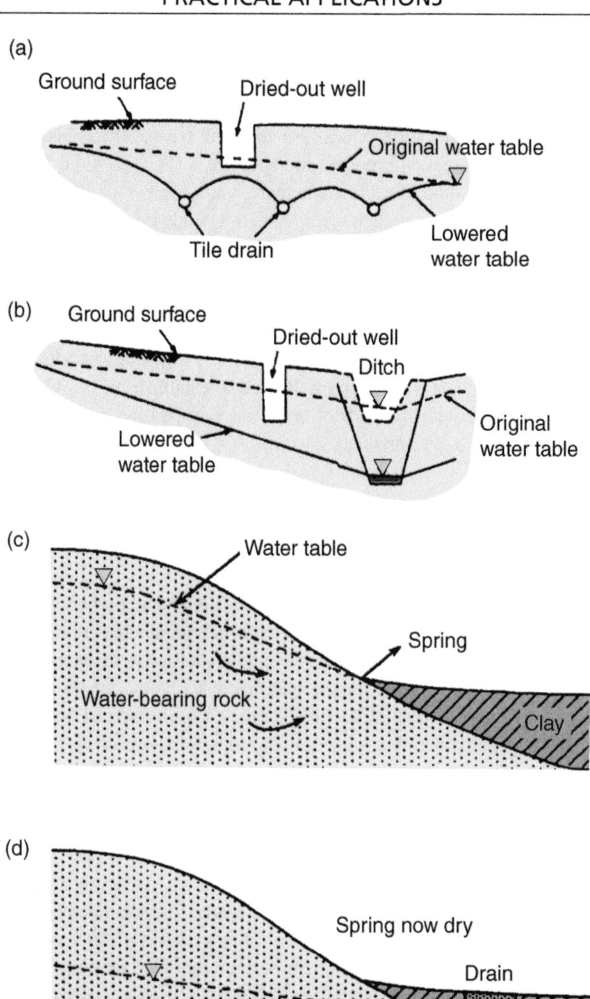

Figure 10.10 *Ditches and drains can provide a route for groundwater discharge that lowers the water table, with potential impacts on local spring and well supplies.*

Rising groundwater levels caused by reduced abstraction will occur over a wide area, contrasting with flooding from a leaking pipe, which will be localised. You may need to investigate the condition of the water pipes and the sewers. It is often difficult to identify the cause and such investigations are often hampered by the 'David and Goliath' effect of private individuals dealing with large organisations such as water companies. You will need to be both resourceful in your hydrogeological assessment and in dealing with the people and organisations involved and persistent in not taking 'No' for an answer!

10.9 Soakaways

Urban drainage has developed since the mid-1990s from an engineering discipline concerned with public health into a much wider subject that may have a hydrogeological aspect where water is discharged to a soakaway or infiltration ponds, trenches and basins. It has become a subject in its own right, with its own literature, conferences, and specialists. The technique of sustainable drainage systems (SUDS) has a lot more than just whether or not you can drain to a soakaway system. SUDS design is influenced by many diverse factors which include the quality of urban runoff, re-use of grey water in buildings, climate change, and developments in paving-over front gardens.

A sustainable drainage system will contain sufficient storage to accommodate storms or large inputs of water and also appropriate treatment so that the discharge to either surface water or groundwater can be slowed down and managed. It is likely that you will become involved as a hydrogeological specialist in a civil engineering team rather than being responsible for the scheme as a whole. On the hydrogeological side, you will need to establish the geology and groundwater conditions, planning the investigation as in Section 2.9, so that you can develop a conceptual understanding of the groundwater system involved. If you need to drill boreholes to monitor the groundwater levels or look at the geology, Section 9.1 will give you guidance. One thing to bear in mind is that, in general, it is far more difficult to get water to flow into the ground than it is to abstract it.

10.10 Investigating Wetland Hydrology

The key characteristic of wetlands is that they are entirely dependent on being saturated with water for all or the greater part of the year. These conditions form habitats that are critical for many plant, insect, and bird species. Large areas of wetlands have been drained in many countries, making wetland habitats rare and, consequently, important to nature conservation. The European Commission's Habitats Directive (Council Directive 92/43/EEC on the Conservation of Natural Habitats and of Wild Fauna and Flora) requires member states to take measures to protect all endangered habitats that may be under threat from a wide range of human activities, including water abstraction, waste disposal, contamination from industrial processes, mining and quarrying, land drainage, and construction works. As a result, the hydrogeology of a large number of wetlands has been investigated across Europe, no doubt with even more to be completed in the future.

There are two basic forms of wetland. The first depends on surface water runoff, where low-permeability materials form a shallow basin that retains water. The second depends on groundwater, where a permanently shallow water table forms areas of open water and waterlogged soils. The first type can occur anywhere where there is sufficient surface runoff to cause ponding. The second is frequently found along river valleys, generally in close proximity to the watercourse. Shallow water tables also occur in other areas such as coastal dunes where seasonal areas of open water are formed along the valleys between the dunes when the water table is high. Figure 10.11 shows simplified examples of how wetlands are formed. The first example (Figure 10.11a) is a wetland that is entirely dependent on surface and rainfall inputs. It probably started as a shallow depression in a clay surface such as a boulder clay area left by retreating ice. In the second example (Figure 10.11b), springs from a shallow aquifer feed into a low-lying area that is underlain by low-permeability clay. Initially, the area would comprise ponds or lakes that eventually fill with decaying vegetation to form a wetland or bog. In the third example (Figure 10.11c), upward seepage from the regional groundwater flow maintains the wetland. In these three examples there will also be a rainfall input, although it is less important than the groundwater inflows.

A sad fact that has been discovered by researchers into wetland systems is that once any drainage has been added into the wetland it starts a process of erosion that is impossible to stop; it is hoped however, that it can be slowed down.

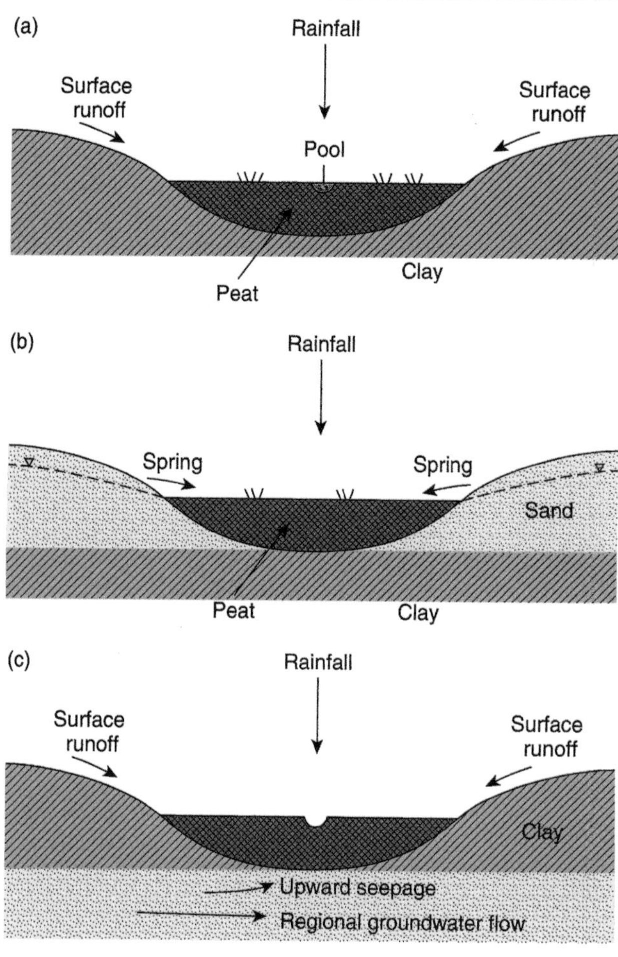

Figure 10.11 *These sketches show how wetlands are formed and rely on a constant water supply.*

The key to protecting and managing a wetland is to understand the surface water and groundwater processes that have caused it to form. Your investigation needs to define the geology of the site and requires piezometers to monitor groundwater levels close to and within the site, and monitor surface water overflows from the site so that a water budget can be quantified. It is likely that your investigation of the wetland will be paralleled by an investigation of the activities that have potential impacts. In most cases, it is more effective and better value to run the investigations together.

Appendix A

GOOD WORKING PRACTICE

Hydrogeological fieldwork involves potential hazards that you must identify before you go out into the field so that you can avoid taking unnecessary risks. This brief note can only alert you in a general way as to what constitutes a potential hazard. It is up to *you* to ensure that you have the proper equipment and training in first aid and dealing with emergencies, and that you follow appropriate safe practices in carrying out your field measurements. Remember that *hazards* cover a wide range of dangers that could affect you, your colleagues helping you in the field, or members of the public who happen to be passing by.

A1.1 Safety Codes

- Most organisations involved in hydrological fieldwork have their safety codes. Make sure that you read and understand them before setting out. If necessary, borrow a safety code from the local water company or similar organisation.
- Leave details of your route and work locations with an appropriate person at your base, with approximate times of your return and details of what action should be taken if you fail to turn up. Do not forget to check in – whatever time you get back – to prevent abortive search and rescue operations.
- Check the local weather forecast on the television, radio or internet information service, and avoid working in severe weather conditions. Poor conditions will reduce the amount of work that you can complete, and by putting yourself at risk you are placing an unnecessary strain on rescue services.
- A great deal of work is likely to be carried out in rural areas, so always adhere to the code set out in Table A.1. Set a good example and try to fit in with the life and work in the countryside. By creating a good impression and developing good relationships with the landowners in your study area you will get much more co-operation.

A1.2 Safety Clothing and Equipment

- Always ensure that your footwear and clothing is suited to the weather, the terrain, and the task you are undertaking. Remain warm and dry in cold areas and avoid overheating in hot weather. Carry spare clothing such as a sweater, waterproof clothing, or sun hat.
- Good quality walking boots give a good grip on rough ground and protect ankles – important where there may be poisonous snakes, scorpions, or spiders.
- On construction sites, in quarries, and near rock faces, wear safety boots and a helmet. Other items of personal safety equipment include a high-visibility jacket, gloves, goggles, and ear protectors for when you are near noisy machines.
- Your personal emergency equipment should include a first-aid kit, mobile phone (cell phone), whistle or flashlight for emergency signals, spare batteries, matches sealed in a waterproof plastic bag, and a survival bag or aluminium foil survival blanket.
- Carry emergency rations, especially bottled water and a high-energy food such as sugar and glucose bars, glucose tablets, or one of the many dried-fruit mixtures recommended in mountain safety books.

Field Hydrogeology, Fifth Edition. Rick Brassington.

Table A.1 *Code for minimum impact of fieldwork on the countryside.*

Risk of fire
Control any fires you may need for cooking or warmth and fully extinguish them. If you smoke, take care not to drop burning matches, cigarette ends, or pipe ash.

Livestock
Close all gates, even if you find them open, to prevent livestock from wandering. Do not force your way through fences, walls or hedges, as any damage could allow animals to escape. Do not chase animals; they may injure themselves, or could turn on you and cause you injury. If you take a dog with you, prevent it from chasing or frightening the animals. Loud noise, the playing of radios or MP3 players will also disturb animals.

Crops
Do not walk through crops – even grass – unless you have express permission of the owner. Stick to recognised footpaths or the edges of fields.

Litter
Do not drop litter of any sort. Take it home with you, or back to your base. When out for several days, acceptable alternatives are burning (but take care with fires) or burial.

Protect water supplies
Do not let any material enter a water supply. Take the precautions listed in this Appendix.

Wildlife
Do not disturb wild animals or birds. Do not pick wild flowers or damage trees or other plants. Some wild animals can be dangerous, so find out about any that may be in your area. Do not eat wild plants or berries unless you know that they are not poisonous.

Country roads
Country roads have special dangers, such as blind corners, high banks, animals, farm vehicles or machinery, and often poor surfaces. Take care when driving. If walking, face oncoming traffic.

General
Set a good example and try to fit in with the life and work in the countryside. This effort is especially important when you are working in a foreign country. Not only is this behaviour basically good manners, but also it will be invaluable in developing good relationships with the landowners in your study area.

A1.3 Distress Signals

- Before relying on your mobile phone (cell phone) check that the network covers the area. In remote places you may need a satellite phone instead. Check that batteries are fully charged. If you use a pay-as-you-go service make sure you have money in your account!
- The accepted emergency distress signal is six blasts on a whistle or six flashes with a mirror or torch, repeated at 1-minute intervals. Rescuers reply with only three blasts or flashes, again repeated at 1-minute intervals that are intended to prevent rescue parties from homing in on each other.

A1.4 Exposure or Hypothermia

- Everyone working in temperate or cold climates and in mountainous areas is vulnerable to hypothermia, or deep-body cooling.
- Getting wet will reduce body temperatures and can bring on hypothermia.
- Learn to recognise the symptoms, both in yourself and in your companions. Initially these are white or pale complexion, violent shivering, and complaints of feeling cold, and later judgement will be impaired and abnormal or irrational behaviour may be displayed – slurring of speech, disturbed vision, stumbling and falling may also occur.

- Get out of the wind and put on dry clothes or windproof materials such as a survival bag. Keep warm. Have a hot sugary or glucose drink – but *avoid alcohol as it can be fatal!*
- Never underestimate the seriousness of hypothermia. Death can occur within one hour of the onset of symptoms. Seek immediate help.

A1.5 Heat Exhaustion

- Heat exhaustion is due mostly to dehydration (water loss) which causes shock, with a rapid pulse, cold clammy skin, thirst, fatigue, and giddiness. Get into a cool, shady place. Restore the water and salt balance by drinking lots and lots of cool water, containing half a teaspoon of common salt (NaCl) per pint. This is approximately equivalent to 5 mg of salt in 1 L of water.
- Avoid working during the hottest part of the day, sit in the shade instead! Do not remove your clothes, as keeping them on prevents the sweat evaporating too fast. Wear light-coloured clothes because they reflect the heat. Eat little and often, as the digestion of large meals uses up a lot of water. Drink often, but boil the water first or use purification tablets, and do *not* drink seawater.

A1.6 Working near Wells, Boreholes and Monitoring Piezometers

- Main risks are falling into a well or large-diameter borehole, and problems caused by the generation of gases.
- Never enter a confined space unless you are properly trained and have carried out tests to show the atmosphere is safe.
- Gases that may flow from an open well or borehole include carbon dioxide, which can cause suffocation; hydrogen sulphide, which is highly toxic; methane, which has inherent dangers of explosion; and carbon monoxide which is colourless, odourless, tasteless and slightly denser than air. It is also highly explosive
- Removing heavy covers can trap hands or feet and cause back-strain. Chambers may be an attractive home to snakes or poisonous insects.
- Reduce the size of the opening to less than 350 mm, or wear a safety-harness with a lifeline no longer than 3 m attached to a secure point or fastened to at least two 'top men'.
- If you are going to use a ladder or step-iron, make sure that they are secure before you trust your weight on them.
- Examine cover plates, staging, and so on before you carry out any work over chambers or wells. Beware of corroded metalwork or rotten and/or slippery timber that may be rotten underneath.

A1.7 Hygiene Precautions for Water Supplies

- Main risks are allowing material to fall into the supply or introducing contaminants on dirty equipment.
- Brush the cover clean and remove dirt lodged in the frame. Avoid knocking dirt into the well or catch pit when removing the cover.
- Disinfect all equipment, including the probe and tape of dippers, depth samplers, and sample pumps.
- Dilute 10 mL of domestic bleach (e.g. Domestos) in 5 L of water to make a chlorine solution to clean your equipment. Do not splash the bleach on your skin or in your eyes, washing any off as quickly as possible using lots of water. Scrub the equipment in the bleach solution and leave it to soak for an hour or two. Alternatively, use a solution of peracetic acid (CH_3CO_3H) diluted to about 5%.
- Do *not* use 'pine' disinfectants, as they are based on phenols, which will contaminate the water supply, causing taste problems that will last for a very long time.
- Never use equipment on drinking water supplies that has been used for contaminant work.

A1.8 Trial Pits

- The main risks are injuries caused by collapsing pits.
- Never enter an unsupported excavation that is greater than 1.25 m deep.
- Waterlogged pits are unstable, so beware when approaching an excavation.
- Backfill trial pits as soon as possible, or fence them off and clearly mark if left unattended.

A1.9 Electrical Equipment

- The main risks are electrocution or being caught in moving parts.
- Use rechargeable 'cable-less' power tools where possible or tools that will run on a low voltage supply (110 volts).
- Use a circuit breaker for all mains connections.
- Beware of sparks if you are working anywhere near explosive gases.
- Make sure that all generators are properly earthed.

A1.10 Filling Fuel Tanks

- The main risks are explosions, fire, or causing contamination by spillage.
- Use only approved containers to store and carry the fuel. Secure containers during transport. Store them inside a bunded area or a second tank. Make sure that there is no smoking or use of mobile phones or electrical equipment in the vicinity of fuel stores.
- Use a plastic sheet to catch any spillage when you are filling the fuel tank and keep absorbent materials handy.
- Switch off engines during the filling operation and avoid fuel coming into contact with hot parts of the engine.

A1.11 Waste Disposal Sites

- The main risks are those associated with busy sites and coming into contact with toxic materials.
- Wear a high-visibility jacket over your protective overalls and wear rubber boots and rubber gloves when handling water samples taken from, on, or near a waste-disposal site.
- Avoid breathing harmful vapours or gases and avoid physical contact with noxious chemicals either as solids or dissolved in groundwater.
- Methane, hydrogen sulphide, and carbon dioxide are all produced by the decomposition of wastes. Be vigilant for these gases accumulating in boreholes and chambers.
- Avoid leachate coming in contact with your skin – some toxic chemicals can be absorbed into the body through the skin.
- Take care when transporting samples to avoid contact in the event of an accident.
- Have regular anti-tetanus and typhoid injections.
- Contact your medical practitioner in the event of *any* illness and advise him or her of the types of waste you have contacted.

A1.12 Stream Flow Measurement

- The main risks are drowning and coming into contact with waterborne diseases.
- Check the stability of the banks before starting work so that you don't fall in even before you start.
- Do not wade in fast-flowing streams or where the bed is uneven. Remember that water can appear deceptively shallow at times.

- Gauge in teams of two people, with one person remaining on the bank with an emergency life-line ready at all times.
- Avoid direct skin contact with the water if you have any cuts or grazes because of the risk of contracting leptospiral jaundice.

  Report the possibility of leptospiral jaundice to your medical practitioner in the event of *any* illness. This is a very serious condition and is difficult to diagnose, as the symptoms can easily be confused with more common infections.

Appendix B

CONVERSION FACTORS

Temperature is usually measured in degrees Celsius (also known as Centigrade) or degrees Fahrenheit. Water freezes at 0°C or 32°F, and boils at 100°C or 212°F. To convert a value in Celsius into Fahrenheit, multiply by nine and divide by five then add 32. To convert a value in Fahrenheit into Celsius, subtract 32 then multiply by five and divide by nine.

Table B.1 *Length (SI unit: metre, m).*

	m	ft	in
1 m	1.000	3.281	39.37
1 ft	3.048×10^{-1}	1.000	12
1 in	2.54×10^{-2}	8.333×10^{-2}	1.000

Table B.2 *Area (SI unit: square metre, m^2).*

	m^2	ft^2	acre	hectare
1 m^2	1.000	10.764	2.471×10^{-4}	1.0×10^{-4}
1 ft^2	9.290×10^{-2}	1.000	2.296×10^{-5}	9.29×10^{-6}
1 acre	4.047×10^{2}	4.356×10^{4}	1.000	4.047×10^{-1}
1 hectare	1.0×10^{4}	1.076×10^{5}	2.471	1.000

Table B.3 *Volume (SI unit: cubic metre, m^3).*

	m^3	litre	Imp gal	US gal	ft^3
1 m^3	1.000	1.000×10^{3}	2.20×10^{2}	2.642×10^{2}	35.32
1 litre	1.000×10^{-3}	1.000	0.220	0.2642	3.532×10^{-2}
1 imp gal	4.546×10^{-3}	4.546	1.000	1.20	1.605×10^{-1}
1 US gal	3.785×10^{-3}	3.785	8.327×10^{-1}	1.000	1.337×10^{-1}
1 ft^3	2.832×10^{-2}	28.32	6.229	7.480	1.000

Field Hydrogeology, Fifth Edition. Rick Brassington.

Table B.4 *Time (SI unit: second, s).*

	s	min	h	d
1 s	1.000	1.667×10^{-4}	2.777×10^{-4}	1.157×10^{-5}
1 min	60	1.000	1.667×10^{-2}	6.944×10^{-4}
1 h	3.6×10^{3}	60	1.000	4.167×10^{-2}
1 d	8.640×10^{4}	1.440×10^{3}	24.00	1.000

Table B.5 *Discharge (SI unit: cubic metre per second, $m^3\ s^{-1}$).*

	$m^3\ s^{-1}$	$m^3\ d^{-1}$	$L\ s^{-1}$	Imp gal d^{-1}	US gal d^{-1}	$ft^3\ s^{-1}$
1 $m^3\ s^{-1}$	1.000	8.640×10^{4}	1.000×10^{3}	1.901×10^{7}	2.282×10^{7}	35.315
1 $m^3\ d^{-1}$	1.157×10^{-5}	1.000	1.157×10^{-2}	2.200×10^{2}	2.642×10^{2}	4.087×10^{-4}
1 $L\ s^{-1}$	1.000×10^{-3}	86.40	1.000	1.901×10^{4}	2.282×10^{4}	3.531×10^{-2}
1 Imp gal d^{-1}	5.262×10^{-8}	4.546×10^{-3}	5.262×10^{-5}	1.000	1.201	1.858×10^{-6}
1 US gal d^{-1}	4.381×10^{-8}	3.785×10^{-3}	4.381×10^{-5}	8.327×10^{-1}	1.000	1.547×10^{-6}
1 $ft^3\ s^{-1}$	2.832×10^{-2}	2.447×10^{3}	28.32	5.382×10^{5}	6.464×10^{5}	1.000

Table B.6 *Hydraulic conductivity (SI unit: cubic metre per second per square metre, reduced to $m\ s^{-1}$).*

	$m\ s^{-1}$	$m\ d^{-1}$	Imp gal $d^{-1}\ ft^{-2}$	US gal $d^{-1}\ ft^{-2}$	$ft\ s^{-1}$
1 $m\ s^{-1}$	1.000	8.640×10^{4}	1.766×10^{6}	2.121×10^{6}	3.281
1 $m\ d^{-1}$	1.157×10^{-5}	1.000	20.44	24.54	3.797×10^{-5}
1 Imp gal $d^{-1}\ ft^{-2}$	5.663×10^{-7}	4.893×10^{-2}	1.000	1.201	1.858×10^{-6}
1 US gal $d^{-1}\ ft^{-2}$	4.716×10^{-7}	4.075×10^{-2}	8.327×10^{-1}	1.000	1.547×10^{-6}
1 $ft\ s^{-1}$	3.048×10^{-1}	2.663×10^{4}	5.382×10^{5}	6.463×10^{5}	1.000

Table B.7 *Transmissivity (SI unit: metres × cubic metre per second per square metre per metre, reduced to $m^2\ s^{-1}$).*

	$m^2\ s^{-1}$	$m^2\ d^{-1}$	Imp gal $d^{-1}\ ft^{-1}$	US gal $d^{-1}\ ft^{-1}$	$ft^2\ s^{-1}$
1 $m^2\ s^{-1}$	1.000	8.640×10^{4}	5.793×10^{6}	6.957×10^{6}	10.76
1 $m^2\ d^{-1}$	1.157×10^{-5}	1.000	67.05	80.52	1.246×10^{-4}
1 Imp gal $d^{-1}\ ft^{-1}$	1.726×10^{-7}	1.491×10^{-2}	1.000	1.201	1.858×10^{-6}
1 US gal $d^{-1}\ ft^{-1}$	1.437×10^{-7}	1.242×10^{-2}	8.327×10^{-1}	1.000	1.547×10^{-6}
1 $ft^2\ s^{-1}$	9.29×10^{-2}	8.027×10^{3}	5.382×10^{5}	6.463×10^{5}	1.000

REFERENCES

Banks, D. (2012) *An Introduction to Thermogeology: Ground Source Heating and Cooling*, 2nd edn, John Wiley & Sons, Chichester.

Bannister, A., Raymond, S. and Baker, R. (1998) *Surveying*, 7th edn, Longman, London.

Bedient, P.B., Rifia, H.S. and Newell, C.J. (1999) *Ground Water Contamination – Transport and Remediation*, Prentice-Hall, Englewood Cliffs, NJ.

Beesley, K. (1986) Downhole geophysics, in *Groundwater: Occurrence, Development and Protection. Water Practice Manual No. 5* (ed. T.W. Brandon), Institution of Water Engineers and Scientists, London, pp. 315–352.

Bierschenk, W.H. (1964) Determining well efficiency by multiple step-drawdown tests. *International Association of Scientific Hydrology*, **64**, 493–507.

Brandon, T.W. (ed.) (1986) *Groundwater: Occurrence, Development and Protection* Water Practice Manual No. 5, Institution of Water Engineers and Scientists, London.

Brassington, F.C. (1992) Measurements of head variations within observation boreholes and their implications for groundwater monitoring. *Journal of the Institution of Water and Environmental Management*, **6**, 91–100.

Brassington, F.C. (2014) Interpretation of groundwater data in the vicinity of a long established sandstone quarry. *Proceedings of the 17th Extractive Industry Geology Conference*, Edge Hill, 5–8 September. www.eigconferences.com.

Brassington, F.C. and Preene, M. (2003) The design, construction and testing of a horizontal well-point as a water source in a dune sands aquifer. *Quarterly Journal of Engineering Geology and Hydrogeology*, **36**, 355–366.

Brassington, F.C. and Taylor, R. (2012) A comparison of field methods used to define saline–fresh groundwater interfaces at two sites in North West England. *Quarterly Journal of Engineering Geology and Hydrogeology*, **45**, 173–181.

Brassington, F.C. and Walthall, S. (1985) Field techniques using borehole packers in hydrogeological investigations. *Quarterly Journal of Engineering Geology*, **18**, 181–193.

Brassington, F.C. and Younger, P.L. (2010) A proposed framework for hydrogeological conceptual modelling. *Water and Environment Journal*, **24**, 261–273.

Brassington, R. (1995) *Finding Water*, 2nd edn, John Wiley & Sons, Chichester.

Busenberg, E. and Plummer, L.N. (1997) Use of sulfur hexafluoride as a dating tool and as a tracer of igneous and volcanic fluids in ground water. *Geological Society of America, Salt Lake City, 1997 Annual Meeting, Abstracts with Programs*, **29** (6), A-78.

Cashman, P.M. and Preene, M. (2013) *Groundwater Lowering in Construction: A Practical Guide*, 2nd edn, Spon, London.

Chilton, P.J. and Foster, S.S.D. (1991) Control of ground-water nitrate pollution in Britain by land-use change, in *Nitrate Contamination*. NATO ASI Series Volume G30 (eds I. Bogárdi and R.D. Kuzelka), Springer-Verlag, New York, pp. 333–347.

Clark, W.E. (1967) Computing the barometric efficiency of a well. *Journal of the Hydraulics Division, American Society of Civil Engineers*, **93** (HY4), 93–98.

Field Hydrogeology, Fifth Edition. Rick Brassington.

Cooper, H.H. and Jacob, C.E. (1946) A generalized graphical method for evaluating formation constants and summarizing well field history. *Transactions of the American Geophysical Union*, **27**, 526–534.

Davis, S.N., Campbell, D.J., Bentley, H.W. and Flynn, T.J. (1985) *Ground Water Tracers*, National Ground Water Association, Westerville, OH.

Driscoll, F.G. (1986) *Groundwater and Wells*, 2nd edn, Johnson Filtration Systems, St Paul, MN.

Environment Agency (2004) *Assessing the Groundwater Pollution Potential of Cemetery Developments Guidance Booklet*, Environment Agency, Rotherham.

Fetter, C.W. (2018) *Applied Hydrogeology*, 4th edn, Waveland Pr Inc., Long Grove, Illinois.

Fetter, C.W., Boving, T. and Kreamer, D. (2018) *Contaminant Hydrogeology*, 3rd edn, Waveland Pr Inc., Long Grove, Illinois.

Gat, J.R. (1971) Comments on the stable isotope method in regional groundwater investigations. *Water Resources Research*, **73**, 980–993.

Grindley, J. (1969) *The calculation of actual evaporation and soil moisture deficit over specified catchment areas Hydrological Memorandum No. 38*, Meteorological Office, HMSO, London.

Grindley, J. (1970) Estimation and mapping of evaporation. *World Water Balance*, **1** (92), 200–213.

Hazen, A. (1911) Discussion: dams on sand foundations. *Transactions of the American Society of Civil Engineers*, **73**, 199.

Headworth, H.G. (1970) The selection of root constants for the calculation of actual evaporation and infiltration for chalk catchments. *Journal of the Institution of Water Engineers and Scientists*, **24**, 569–574.

Hem, J.D. (1985) *Study and Interpretation of the Chemical Characteristics of Natural Water*, 3rd edn, Geological Survey Water-Supply Paper 2254. US Department of the Interior, Washington, DC.

Henry, W. (1803) Experiments on the quantity of gases absorbed by water at different temperatures and under different pressures. *Philosophical Transactions of the Royal Society*. **93**, 29–274

Hill, R.A. (1940) Geochemical patterns in the Coachella Valley, California. *Transactions of the American Geophysical Union*, **21**, 46–49.

Hinkle, S.R., Shapiro, S.D., Plummer, L.N., Busenberg, E., Widman, P.K., Casile, G.C. and Wayland, J.E. (2010) Estimates of tracer-based piston-flow ages of groundwater from selected sites: national water-quality assessment program, 1992–2005. *Scientific investigations report 2010–5229*. U.S. Department of the Interior & U.S. Geological Survey, 90 pp.

Hough, M.N. and Jones, R.J.A. (1997) The United Kingdom Meteorological Office rainfall and evaporation calculation system: MORECS version 2.0 – an overview. *Hydrology and Earth Systems Science*, **1**, 227–239.

Hubbert, M.K. (1940) The theory of ground-water motion. *Journal of Geology*, **48**, 785–944.

Hvorslev, M.J. (1951) Time lag and soil permeability in groundwater observations *Bulletin No 36, Waterways Experimental Station*, Corps of Engineers, Vicksburg, MI.

Jacob, C.E. (1939) Fluctuations in artesian pressure produced by passing railroad-trains as shown in a well on Long Island, New York. *Transactions of the American Geophysical Union*, **20**, 666–674.

Kruseman, G.P. and de Ridder, N.A. (1994) *Analysis and Evaluation of Pumping Test Data*, 2ndILRI Publication 47 edn, International Institute for Land Reclamation and Development, Wageningen.

Lerner, D.N., Issar, A.S. and Simmers, I. (1990) *Groundwater recharge. A guide to understanding and estimating natural recharge International Contributions to Hydrogeology* **8**, International Association of Hydrogeologists, Heise, Hannover.

Lisle, R.J., Brabham, P. and Barnes, J.W. (2011) *Basic Geological Mapping*, 5th edn, Wiley, Chichester.

Logan, J. (1964) Estimating transmissibility from routine production tests of water wells. *Ground Water*, **2**, 35–37.

Mailvaganam, Y., Ramili, M.Z., Rushton, K.R. and Ong, B.Y. (1993) Groundwater exploitation of a shallow coastal sand aquifer in Sarawak, Malaysia, in *Hydrology of Warm Humid Regions. Proceedings of the Yokohama Symposium*, vol. **216**, IAHS Publication, pp. 451–462.

McDonald, M.G. and Harbaugh, A.W. (1988) *A modular three- dimensional finite- difference ground- water flow model. Techniques of Water-Resources Investigations of the United States Geological Survey, Book 6*, US Geological Survey, Denver.

Meinzer, O.E. (1923) The occurrence of ground water in the United States, with a discussion of principles. *US Geological Survey Water-Supply Paper 489*. US Geological Survey, Denver, CO.

Milsom, J.S. and Eriksen, A. (2011) *Field Geophysics*, 4th edn, Wiley, Chichester.

Misstear, B., Banks, D. and Clark, L. (2017) *Water Wells and Boreholes*, 2nd edn, John Wiley & Sons, Chichester.

Penman, H.L. (1948) Natural evaporation from open water, bare soil and grass. *Proceedings of the Royal Society*, **193**, 120–145.

Penman, H.L. (1950a) Evaporation over the British Isles. *Quarterly Journal of the Royal Meteorological Society*, **96**, 372–383.

Penman, H.L. (1950b) The water balance of the Stour catchment area. *Journal of the Institution of Water Engineers*, **4**, 457–469.

Piper, A.M. (1944) A graphic procedure in the geochemical interpretation of water analyses. *Transactions of the American Geophysical Union*, **25**, 914–928.

Preene, M., Roberts, T.O., Powrie, W. and Dyer, M.R. (2000) *Groundwater Control – Design and Practice* CIRIA Report C515, Construction Industry Research and Information Association, London.

Price, M. and Williams, A.T. (1993) A pumped double-packer system for use in aquifer evaluation and groundwater sampling. *Proceedings of the Institution of Civil Engineers*, **101**, 85–92.

Rushton, K.R. and Brassington, F.C. (2013a) Hydraulic behaviour and regional impact of a horizontal well in a shallow aquifer: example from the Sefton Coast, northwest England (UK). *Hydrogeology Journal*, **21** (5), 1117–1128.

Rushton, K.R. and Brassington, F.C. (2013b) Significance of hydraulic head gradients within horizontal wells in unconfined aquifers of limited saturated thickness. *Journal of Hydrology*, **492**, 281–289.

Rushton, K.R. and Brassington, F.C. (2016) Pumping from unconfined aquifers of limited saturated thickness with reference to wellpoints and horizontal wells. *Hydrogeology Journal*, **24** (2), 335–348.

Rushton, K.R. and Holt, S.M. (1981) Estimating groundwater parameters for large diameter wells. *Ground Water*, **19**, 505–509.

Rushton, K.R., Kawecki, M.W. and Brassington, F.C. (1988) Groundwater model conditions in Liverpool sandstone aquifer. *Journal of the Institution of Water and Environmental Management*, **2**, 67–85.

Schafer, D.C. (1978) Casing storage can affect pumping test data. *Johnson Drillers' Journal*, **50**, 1–5.

Schoeller, H. (1962) *Les Euax Souterraines*, Masson & Cie, Paris, 642 pp.

Selker, J. and Or, D. (2022) *Soil Hydrology and Biophysics*. Licensed under a Creative Commons Attribution-NonCommercial-ShareAlike 4.0 International License. https://open.oregonstate.education/soilhydrologyandbiophysics.

Smith, D.B., Wearn, P.L., Richards, H.J. and Rowe, P.C. (1970) Water movement in the unsaturated zone of high and low permeability strata by measuring natural tritium. *IAEA Symposium on Isotope Hydrology*, Vienna, pp. 73–87.

Stuart, M.E., Ward, R.S., Ascott, M. and Hart, A.J. (2016) *Regulatory practice and transport modelling for nitrate pollution in groundwater*. British Geological Survey Internal Report, OR/16/033.

Theis, C.V. (1935) The relation of the lowering of the piezometric surface and the rate and duration of the discharge of a well using groundwater storage. *Transactions of the American Geophysical Union*. **16**, 519–524

Thiem, G. (1906) *Hydrologische Methoden*, Gebhardt, Leipzig.

Thiessen, A.H. (1911) Precipitation Averages for Large Areas. Monthly Weather Review, **39**, 1082–1089.

Thornthwaite, C.W. (1948) An approach towards a rational classification of climate. *Geographical Review*, **38**, 55–94.

Toll, N.J. and Rasmussen, T.C. (2007) Removal of barometric pressure effects and Earth tides from observed water levels. *Ground Water*, **45**, 101–105.

United Nations (1976) *Groundwater in the Western Hemisphere*, United Nations, New York.

US Department of the Interior Bureau of Reclamation (1995) *Ground Water Manual – A Water Resources Technical Publication*, 2nd edn, US Government Printing Office, Washington, DC.

Walthall, S. (1999) Packer tests in geotechnical engineering, in *Field Testing in Engineering Geology*. (Geological Society Engineering Geology Special Publication), No 6 (eds F.G. Bell, M.G. Culshaw, J.C. Cripps and J.R. Coffey), Geological Society, London, pp. 345–350.

Ward, R.E., Williams, A.T., Barker, J.A. *et al.* (1998) *Groundwater Tracer Tests: A Review and Guidelines for their Use in British Aquifers*, British Geological Survey Technical Report WD/98/19 Hydrogeology Series(R&D Technical Report W160), British Geological Survey, Keyworth.

White, K.E. (1986) Tracing techniques, in *Groundwater: Occurrence, Development and Protection. Water Practice Manual No. 5* (ed. T.W. Brandon), Institution of Water Engineers and Scientists, London, pp. 353–383.

Wilkinson, W.B. and Brassington, F.C. (1991) Rising groundwater levels – an international problem, in *Applied Groundwater Hydrology* (eds R.A. Downing and W.B. Wilkinson), Clarendon, Oxford, pp. 35–53.

Younger, P.L. (1993) Simple generalized methods for estimating aquifer storage parameters. *Quarterly Journal of Engineering Geology*, **26**, 127–135.

Scientific Papers

The following international journals contain papers on a variety of current hydrogeological topics and should be read by practising hydrogeologists.

Ground Water. Published six times per year by the National Water Well Association, Dublin, OH.

Ground Water Monitoring and Remediation. Published quarterly by the Ground Water Publishing Company, Westerville, OH.

Hydrogeology Journal (formerly called *Applied Hydrogeology*). Published six times per year by the International Association of Hydrogeologists.

International Groundwater Technology. Published bi-monthly by National Trade Publications Inc., Latham, NY.

Journal of Hydrology. Published with four issues per volume and four volumes per year by Elsevier Scientific Publishing Company, Amsterdam.

Quarterly Journal of Engineering Geology and Hydrogeology. Published four times per year by the Geological Society, London.

Although not strictly scientific journals in the same sense as the titles listed above, there are three series of publications that contain invaluable information for professional hydrogeologists. These are the following:

International Contributions to Hydrogeology. Published by the International Association of Hydrogeologists, Reading.

Some of the Special Publications of the Geological Society of London, Burlington House, Piccadilly, London W1J 0BG.

Water Supply Papers. Published by the US Geological Survey, Denver, CO.

Further Reading

Anon (1963) Treaty banning nuclear weapon tests in the atmosphere, in outer space and under water. Between the United States, the United Kingdom and the Union of Soviet Socialist Republics.

Brassington, F.C., Lucey, P.A. and Peacock, A.J. (1992) The use of down-hole focused electric logs to investigate saline ground waters. *Quarterly Journal of Engineering Geology*, **25**, 343–350.

Brassington, F.C., Whitter, J.P., Macdonald, R.A. and Dixon, J. (2009) The potential use of hydrogen peroxide in water well rehabilitation. *Water & Environment Journal*, **23**, 69–74.

British Standards Institution (2017) *Hydrometry. Open Channel Flow Measurement Using Thin-Plate Weirs*. BS ISO 1438:2017, British Standards Institution, London.

British Standards Institution (2003) *Hydrometric Determinations. Pumping Tests for Water Wells. Considerations and Guidelines for Design Performance and Use*. BS ISO 14686:2003, British Standards Institution, London.

Chapelle, F.H. (2001) *Ground-Water Microbiology and Geochemistry*, 2nd edn, Wiley, New York.

Clark, L. (1977) The analysis and planning of step drawdown tests. *Quarterly Journal of Engineering Geology*, **10**, 125–143.

Clark, L. (1992) *Methodology for Monitoring and Sampling Groundwater* R&D Note 126, Environment Agency, Bristol.

Detay, M. (1997) *Water Wells: Implementation, Maintenance and Restoration*, Wiley, Chichester.

Domenico, P.A. and Schwartz, F.W. (1997) *Physical and Chemical Hydrogeology*, 2nd edn, Wiley, New York.

Downing, R.A. and Wilkinson, W.B. (eds) (1991) *Applied Groundwater Hydrology*, Clarendon, Oxford.

Field, M. (1984) The Meteorological Office rainfall and evaporation calculation system – MORECS. *Agricultural Water Management*, **6**, 297–306.

Freeze, R.A. and Cherry, J.A. (1979) *Groundwater*, Prentice-Hall, Englewood Cliffs, NJ.

Head, K.H. (1982) *Manual of Soil Laboratory Testing (Vols 1 and 2)*, Pentech, London.

Hiscock, K. and Bense, V. (2014) *Hydrogeology: Principles and Practice*, 2nd edn, Blackwell Publishing, Oxford.

Jacob, C.E. (1947) Drawdown test to determine effective radius of artesian well. *Transactions of the American Geophysical Union*, **112**, 1047–1070.

Price, M. (1996) *Introducing Groundwater*, 2nd edn, Chapman and Hall, London.

Reeves, M.J. (1986) Well tests, in *Groundwater Occurrence, Development and Protection* (ed. T.W. Brandon), Institution of Water Engineers and Scientists, London, pp. 135–188.

Rushton, K.R. (2003) *Groundwater Hydrology: Conceptual and Computational Models*, John Wiley & Sons, Chichester.

Todd, D.K. and Hayes, L.W. (2004) *Ground Water Hydrology*, 3rd edn, Wiley, New York.

United States Environmental Protection Agency (1986) *RCRA Groundwater Monitoring Technical Enforcement Guidance Document* OSWER-9950.1, US Government Printing Office, Washington, DC.

Wilson, E.M. (1990) *Engineering Hydrology*, 4th edn, Macmillan, London.

Younger, P.L. (2007) *Groundwater in the Environment: An Introduction*, Blackwell Publishing, Oxford.

INDEX

Page locators in **bold** indicate tables. Page locators in *italics* indicate figures. This index uses letter-by-letter alphabetization.

Field Hydrogeology, Fifth Edition. Rick Brassington.

Printed and bound by CPI Group (UK) Ltd, Croydon, CR0 4YY

25/08/2023

08104965-0001